Design Rules for Permanent Magnet
Synchronous Machine with Tooth Coil Winding Arrangement

Design Rules for Permanent Magnet Synchronous Machine with Tooth Coil Winding Arrangement

Von der Fakultät für Elektrotechnik, Informationstechnik, Physik

der Technischen Universität Carolo-Wilhelmina zu Braunschweig

zur Erlangung des Grades eines Doktors

der Ingenieurwissenschaften (Dr.-Ing.)

genehmigte Dissertation

von:	Ahamed Bilal Asaf Ali
aus:	Madurai, India
eingereicht am:	06.01.2014
mündliche Prüfung am:	04.07.2014
1. Referent:	Prof. Dr.-Ing. Wolf-Rüdiger Canders
2. Referent:	Prof. Dr.-Ing. Michael Kurrat
Druckjahr:	2014

Bibliografische Information der Deutschen Nationalbibliothek

Die Deutsche Nationalbibliothek verzeichnet diese Publikation in der Deutschen Nationalbibliografie; detaillierte bibliografische Daten sind im Internet über http://dnb.d-nb.de abrufbar.

1. Aufl. - Göttingen: Cuvillier, 2014

Zugl.: (TU) Braunschweig, Univ., Diss., 2014

Fakultät für Elektrotechnik, Informationstechnik, Physik

© CUVILLIER VERLAG, Göttingen 2014
Nonnenstieg 8, 37075 Göttingen
Telefon: 0551-54724-0
Telefax: 0551-54724-21
www.cuvillier.de

1. Auflage, 2014
Gedruckt auf umweltfreundlichem, säurefreiem Papier aus nachhaltiger Forstwirtschaft

ISBN 978-3-95404-770-3
eISBN 978-3-7369-4770-2

Acknowledgements

The dissertation work was done during my career as Scientific Research Assistant at Institute for Electrical Machines, Traction and Drives, IMAB, TU-Braunschweig.

First, I would like to express my sincere gratitude to Prof. Dr.-Ing. Wolf.-Rüdiger Canders for giving me the opportunity to involve me in industrial project and later provided me very good topic for my Dissertation. His valuable feedback on my work contributed greatly to this dissertation work. My special thanks goes to Prof. Dr.-Ing. Markus Henke, Director of IMAB for accepting the role as head of the examination committee. My foremost thanks also goes to Prof. Dr.-Ing. Michael Kurrat who showed interest in my work as well as for being my co-examiner.

I thank Dr.-Ing. Helmut Mosebach for his encouragement and insight suggestions that helped to shape my research skills. My special thanks goes to the industrial partner AS Drives & Services GmbH for the financial support and for the nice team work with R&D Manager Mr. Andre Jagodowski who provided measurement results to validate my work. I thank Dr.-Ing. Nicolai Lescow from IMAB who took responsibility for the inverter part of the test bench during this project and his valuable discussion.

I would like to acknowledge all the colleagues at IMAB who accompanied me by making the work very enjoyable and memorable. I thank Dr.-Ing. Dennis Hülsmann for German abstract correction. I thank M.Sc. Mang Cai for his timely help. My special thanks goes to Dipl.-Ing Christoph Löffler, Frau Barbara Tiedge and Dipl.-Ing Anna-lena Menn for all their support.

I thank the students Yury, Cong, Eduardo and Jue for their support at IMAB. I would like to thank my friends Vijayanantham, Nanda, Mehe, Bilal, Abbas, Ikram and Shamsul for their motivation and support. I also thank Prof. Dr. Damir Zarko for his friendly discussion.

I am thankful to all my family members and especially my mother V.K.H. Abdul Kathoon for giving me moral support. I also thank my wife's family members. Finally, I am grateful to my wife Thasneem, my daughter Fadilah and my son Sakeen Ibrahim for their patience and who made me happy during these times.

Waldshut-Tiengen, July 2014

Ahamed Bilal Asaf Ali

Abstract

This thesis presents the generic rules for permanent magnet synchronous machine with tooth coil winding arrangement. The generic rules concentrates on minimized cogging torque and torque ripple. In this dissertation work, paper-mill is selected as an application. Initially, literature research is provided about the existing design rules. Moreover, rough comparison of direct drive and drive with gears is explained using simplified approach. The geometries considered in this thesis are two different tooth coil winding arrangements and three different rotor types.

The armature flux density and the permanent magnet flux density are derived analytically. The permeance function which is also the source for cogging torque and torque ripple is discussed. The cogging torque and torque ripple are derived analytically using the stator field, rotor field and the permeance functions. The detailed torque analysis of all the considered geometries are performed in Finite Element Method (FEM) for different slot opening and magnet pole coverage. The 2D harmonics analysis approach is used to predict the sources of the harmonics from the numerical calculation. The harmonic sources are reconstructed and validated with the pulsating torque obtained directly from the FEM. The results from the reconstructed curve and the curve which directly obtained from the FEM are in good agreement.

The harmonic sources of pulsating torque are also validated with prototype for a geometry. The investigations on pulsating torque are extended to other operating points such as field weakening and half load condition. Alternative field weakening method is also examined for pulsating torque. Moreover, optimization tool is also used in this thesis to find the minimum cogging torque. In all the cases, the corresponding sources of harmonics for pulsating torque are discussed.

Finally, a generic design rule is suggested for permanent magnet synchronous machine (PMSM). In addition, simplified design rules are suggested to have quick design approach for PMSM. Since manufacturing plays an vital role for a product, this thesis also shows design guidelines from manufacturing view point.

Kurzfassung

In dieser Arbeit wird die generische Gestaltung von dreisträngigen, permanent-magneterregten Synchronmaschinen mit Zahnspulenwicklungsanordnung diskutiert. Dabei liegt der Fokus auf denjenigen generischen Regeln, die zur Minimierung des Rastmomentenverlaufs und der Drehmomentwelligkeit genutzt werden können. Es werden beispielhaft Antriebe in einer Papiermühle-Anwendung betrachtet.

Zuerst wird eine Literaturrecherche über die bestehenden Entwurfsregeln durchgeführt. Außerdem wird ein grober Vergleich zwischen einem Direktantrieb und einem Antrieb mit Getriebe dargestellt. Die Untersuchungen in dieser Arbeit beschränken sich auf zwei verschiedene Zahnspulenwicklungsanordnungen und drei unterschiedliche Rotortypen.

Sowohl die Anker- als auch die Erregerfeldverteilung werden mit Hilfe eines zweidimensionalen , analytischen Ansatzes beschrieben. Unter Berücksichtigung der magnetischen Leitwertfunktion des Luftspaltbereichs lassen sich daraus Werte für das Rastmoment und die Drehmomentwelligkeit ableiten. Die vergleichenden, detaillierten Drehmomentberechnungen der betrachteten Geometrien werden mittels FE-Analysen für verschiedene Nutöffnungen und Polbedeckungen durchgeführt. Der analytisch harmonische Ansatz wird verwendet, um die harmonischen Quellen der numerischen Berechnung zu identifizieren. Diese Quellen werden rekonstruiert und mit dem pulsierenden Drehmoment der FE-Analyse verglichen. Die Ergebnisse zeigen eine gute Übereinstimmung zwischen der rekonstruierten Kurven und den Kurven aus den FE-Analysen.

Die ausführliche Drehmomentanalyse der betrachteten Geometrie werden im FEM für verschiedene Nutöffnung und Polbedeckung durchgeführt . Jedoch werden einige Parameter konstant gehalten für das Drehmoment Analyse. 2D Harmonischen Analyse Ansatz wird verwendet, um die harmonischen Quellen aus der numerischen Berechnung vorherzusagen. Die harmonischen Quellen werden rekonstruiert und mit dem pulsierenden Drehmoment direkt von FEM-Analyse validiert. Die Ergebnisse zeigen eine gute Übereinstimmung zwischen der rekonstruierten Kurve und der Kurve aus der FEM-Analyse.

Die harmonischen Quellen des pulsierendes Drehmoments wurden mit einem Prototyp entsprechend einer Geometrie validiert. Weiterhin lassen sich die Untersuchungen zum pulsierenden Drehmoment auch auf andere Arbeitspunkte, wie Feldschwächung oder Teillast, erweitern. Eine alternative Methode zur Feldschwächung wird ebenfalls hinsichtlich des Rastmomentenverlaufes untersucht. Ferner wird das in dieser Arbeit präsentierte Optimierungs-Tool zum Auffinden des minimalen Rastmoments verwendet, wobei in allen Fällen die entsprechenden Quellen von Harmonischen des pulsierenden Drehmomentes zur Verfügung gestellt werden.

Abschließend wird eine generische Entwurfsregel für permanentmagneterregte Synchron-maschinen empfohlen, die zusätzlich auch in vereinfachter Form dargeboten wird, um möglichst schnell ein geeignetes Maschinendesign zu erhalten. Da die Herstellungsverfahren eine wichtige Rolle beim Maschinendesign spielen, werden in dieser Arbeit die Entwurfs-regeln auch von diesem Standpunkt aus betrachtet.

Contents

3. General Derivation of Force in PMSM

4. Torque Analysis

1. Introduction and Literature

Generally there are various low-speed drive systems available in the industry. Existing components are replaced with modern technology to have high power density, reduced losses, improved power electronics control, computer based monitoring of the drive and to have reliable and continuous operation of the drive system. Additionally, direct drive allow high acceleration and higher dynamics in production machines e.g. milling industries. The modern drive system can save energy by reducing the energy consumption costs as well as global emission [1]. Thus modernization of industries are necessary due to the economical and environmental aspects. Typical applications of low-speed drives include paper and pulp industry, ship propulsion, wind turbine, etc.

Enormous amount of papers, which is one of the basic necessities in day to day life, is used in different forms by us. These papers are manufactured in bulk volumes by the paper industries during the process of paper making, thus consuming 30 % of the produced renewable and non-renewable energy sources in countries like Germany [2]. There are more than 150 paper manufacturers present in Germany [3]. Therefore, improvements of efficient drive systems for the paper industries are a challenging task which leads to save energy. The current trend in paper industry is to have direct drive solution by reducing the mechanical components and compact high torque direct drive to have more surrounding space. Asynchronous Machine (ASM) drives have been widely used in the paper industry and it still exists commercially. However proper selection of machine will have best choice when chosen in terms of better efficiency, high power density, high control dynamics and reduced space.

The alternative solution to the ASM is permanent magnet synchronous machine (PMSM). PMSM have high dynamics and precise control and ASM has complicated field oriented control strategies. In direct drive the motor is directly coupled to the load and therefore reduction in the mechanical components which is a problem for the drive with gears. The reliability of the transmission in the drive system with gears is less and hence the replacement of the gearbox is earlier than expected [4]. There are different types of PMSM. The rotor can be arranged either inner or outer; the permanent magnets can be placed on the surface or inserted in to the rotor; the stator winding types are either distributed arrangement or fractional slot arrangement. The disadvantage of PMSM is that it is subjected to demagnetization under load and short circuit conditions. The other disadvantages are, it has magnet losses under no-load operation and Joules losses during field weakening which is not the case in ASM. Nevertheless, PMSM with fractional-slot arrangement is a good choice of candidate for low-speed direct drive application due to reduced copper volume which reduces the joules loss and increases the efficiency. It also has reduced production costs and has high power density.

1.1. Design Process of Drive System

The details of the direct-drive are discussed in Chapter 2. The approach from design specification to the prototype development and mass production is shown in the Figure 1.1. The goal

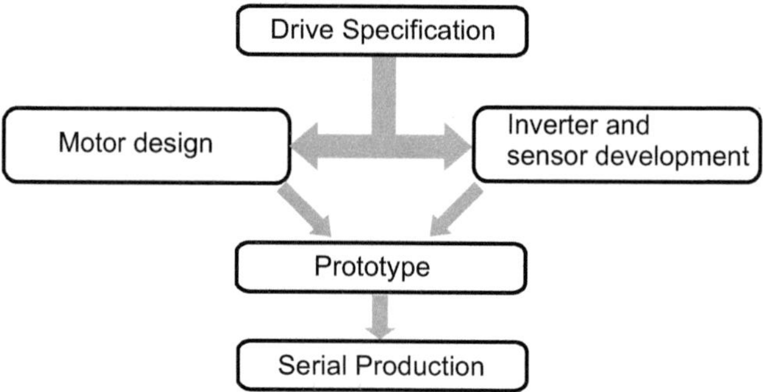

Figure 1.1.: Design of direct drive system.

of drive configuration is to reduce the overall drive dimensions and to improve efficiency. The typical design specifications for paper-mill application are mentioned below.

- A rated torque of 3 $kN.m$ at a rated speed of 200 min^{-1}.

- Range of speed in the constant power region (field weakening) from 200 min^{-1} to 400...500 min^{-1}.

- The DC Link voltage is 690 V.

- The motor outer dimension is restricted to a diameter of 750 mm.

The key component in the electromechanical energy conversion is the electrical machine. There are certain analytical approaches to calculate the geometry dimension parameters of the electrical machines. These approaches are discussed in publications [5] and [6]. Basic design is done analytically with initial selection of force density, magnetic flux density in the magnetic circuit such as air gap, stator and rotor and current density of the stator coil, etc. The dimension of the geometry can be obtained together with the electrical parameters and the power balance of the energy conversion. This geometry is further optimized with the help of finite element method program (FEM). The FEM calculation details are described in detail in the literature [7]. A routine process using software program is done by varying the initial design parameters until the geometry fulfills the given requirement. Next step is to prepare a prototype model and validating the results with FEM results. Usually in prototype the machine is tested for synchronous generated voltage, torque for the given current, mechanical power, efficiency, overload capability, thermal and mechanical durability, short circuit behavior, etc. Once the prototype gives most promising results then the machine is ready for the mass production.

1.2. Objective

When PMSM is selected as a drive motor for low-speed application the next question that arises is about its pulsating torque. Pulsating torque is of two kinds; Cogging torque under no-load operation and Torque ripple under load operation. The methods to reduce the pulsating torque are described in many publications. Few publications are discussed at the end of this chapter. Based on the literature, torque pulsations are minimized either using control methods or by design parameters. If there is negligible pulsation in the PMSM, one can easily turn its rotating part with a free hand. As shown in Figure 1.2, the motor was a prototype with increased torque pulsation and a lever is required for the person to overcome the pulsating torque.
The objective of this dissertation is to develop the generic design rules for permanent magnet synchronous machine for reduced torque pulsation. However, other factors in the existing rules such as losses, average torque, efficiency, machine utilization, mechanical stiffness and economical aspects are also considered to some extent. The work focuses on detailed analysis

of pulsation torque of two fractional slot winding arrangements and three permanent magnet rotor arrangements. The basic design parameters are slot opening, ratio of pole arc to pole

Figure 1.2.: Torque pulsation sense.

pitch (pole coverage). These design parameters are further analyzed by considering different stator winding arrangements and rotor geometries. It should be noted that the skewing method is not taken as a measure to reduce the torque ripple although it is used. In this work, the input current for the analysis of the torque pulsation is considered as sinusoidal. Harmonics due to inverters are not considered because it is not possible to eliminate pulsating torque from design side due to the presence of these harmonic contents. Finally the design rules gives information to design the geometry of a machine. Although in this work, two winding types and three rotor types are considered as examples, these rules should be valid in general because the rules are formed based on the waves in the fields, for example, the selection of geometrical parameters for minimum cogging torque.

1.3. Structure of the Work

Chapter 2 gives an introduction of PMSM in high torque application. This chapter provides information about the drive systems arrangement, difference between drive with gears and without gears and design and working principle of PMSM. Finally, this chapter discusses about considered geometry in this dissertation work.

Chapter 3 gives an introduction to the general derivation of force in PMSM based on Lorentz force and Maxwell's stress tensor. This chapter begins with the study of parasitic torque and permeances in PMSM. The mathematical formulation of the torque pulsation is also explained in this chapter. Analytically derived results are compared with the FEM results in this chapter.

Chapter 4 discusses about detailed analysis of the torque pulsation for the considered geometries. The harmonics which are responsible for the pulsating torque are tabulated. The reconstruction of the torque curve using two dimensional Fast Fourier Transform (2D FFT) is discussed in this chapter.

Chapter 5 describes the measured results and torque pulsation special study. This study includes torque minimization using optimization algorithm and influence of torque pulsation during other operating points. For example torque pulsation is increased during field weakening [4]. This section also discusses about preserving the minimized torque pulsation using alternative field weakening method and torque pulsation during other operating points.

Chapter 6 explains to derive design rules of PMSM for reduced torque pulsation. Finally, the thesis is concluded with the remarks on PMSM rules and future developments. The losses and short circuit analysis are not considered in this work. Basic details of Schwartz-Christoffel transformation, field solutions of current sheet, model analysis of rotor and flux lines and flux density of geometry are provided in the appendix.

1.4. Pulsating Torque Minimization Rules- Literature Discussion

The method to minimize torque pulsations can be classified in to two categories. Design parameters such as slot opening and pole coverage (ratio of pole arc to pole pitch), stator slot and rotor pole combinations, eccentricities of the rotor, permanent magnet magnetization, stator slot shape, permanent magnet shape, auxiliary teeth slots, lamination materials, saturation, number of slots per pole per phase in winding arrangements, manufacturing tolerances and skewing have influence on torque pulsation from the design side. However, optimal design solution is somewhere in the above mentioned design parameters. Alternatively torque pulsation is reduced using control methods (Active torque ripple minimization method). However, there are limitations using control methods. For example, torque ripple in PMSM is a function of rotor angular position. The compensation of torque ripple can be done by modulating the torque producing current component using closed control loop. In this control method, voltage reserve has to be taken in to account for the modulating current. Also, there is also maximum limit for switching frequency of the inverter. Further, during the field weakening, the speed and switching frequency are high which increases the inverter switching losses. Especially when minimizing torque ripple using control methods in field weakening operation, both switching frequency and voltage reserve should be thoroughly considered.

In the following section general rules based on the literature (geometry design ripple minimization method and active ripple minimization method)has been provided for the minimum torque pulsation.

1.4.1. Minimization Method - Design Parameters

Slot opening and pole coverage

The thesis of reference [8] discussed methods to reduce torque ripple by varying slot opening and pole coverage for three different rotor structures of permanent magnet motors with fractional slot winding ($q < 1$) arrangement. However the source of the torque pulsation are not provided. This analysis is also made by several authors in the past decades and it is not quoted in the work. The optimal pole coverage of the magnets are reported in reference [9].

Winding arrangements

Torque pulsation focussed on number of slots per pole per phase (q) is studied in paper [13]. The work from [14] for fractional slot winding investigates various winding arrangements to eliminate the sub-harmonics and harmonics. Magnussen [15] compared the integral slot winding with fractional slot winding.

Air gap field contour

Several methods are reported by influencing the air gap field distribution in order to reduce harmonics magnitude and thus minimizing the torque ripple. In [16] the force oscillations are reduced with shaped magnets. Paper [17] investigates the contour of magnet shape and teeth, curved magnet edge, unequal magnet thickness, introducing slot wedge with small relative permeability and asymmetry arrangement of magnets for an axially excited surface mounted PMSM. The other interesting techniques by using HALBACH type PM and pole shape variation are discussed in the paper [18]. Introducing auxiliary tooth and unequal tooth have reduced cogging torque [10]. In this work the investigation on air gap field closely differentiate the relation between reluctance effect (cogging) and force during load operation due to the super position of field harmonics in the machine.

Stator slot and rotor pole combinations

Significant effects of slot and pole combinations related to torque pulsation are studied in [10] and [8]. Most promising slot-pole combinations are provided. Salminen [8] made analysis for pull-out torque by varying either stator slot or rotor pole by keeping the other geometry parameters as constant. Approaches to find the reduced cogging torque with the help of least common multiple(LCM) are reported in [11] and [8]. Ackermann [12] in his paper also discusses the same (minimum cogging torque) based on the rate of change of surface area of magnets over armature teeth.

Manufacturing defects, Saturation and Eccentricities of the rotor

Additional cogging torque due to manufacturing imperfections are measured and calculated in [19]. Cogging torque effect on PM magnetization faults is explained in [20]. Due to saturation and eccentricity in the machine, the permeances of the magnetic circuit will modify. It generates additional harmonics in the air gap field accounting for pulsating torque and radial forces. The saturation effect and eccentricity effect are studied in [21] and the reparametrization of air gap field due to saturation in [22].

Skewing

Skewing is the common approach to reduce the parasitic torques. This method ideally eliminates the cogging torque but at the expense of the average torque, complication during manufacturing process and increased leakage inductance and stray losses. This approach is discussed in the papers [23], [17] and [24]. This method is not handled in this work.

1.4.2. Minimization Method - Active Control

Harmonic cancellation - Current profiling

Under load, torque ripple minimization can be minimized using additional control method. Optimal current profiling to influence the harmonics is discussed in papers from [18] and [25].

Torque-Position feedback to look-up table

This method is an on-line estimation. The torque versus rotor position information is entered in the look-up table and added to the torque controller to get ripple free reference torque current component as shown in Figure 1.3.

Adaptive instantaneous control

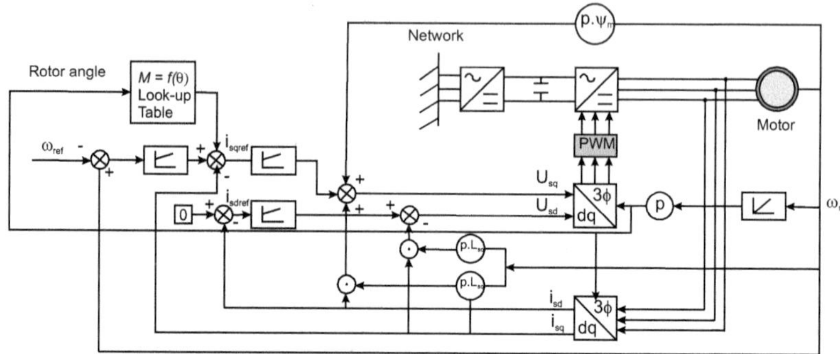

Figure 1.3.: PMSM Torque-Position control

The controller implementation and minimization of torque ripple with the help of adaptive control algorithm is discussed in the paper [26]. This paper also discusses about the experimental results for the controller implementation. The above mentioned solutions to reduce the torque ripple did not discuss about the other operating points. The other operating points are half the rated load and field weakening operation. For industry and automotive applications torque pulsations at these operating points are necessary. Torque pulsation during these operating points is also studied in this work.

Existing rules are summarized in Figure 1.4. In the design methods mentioned in the literature, it ss possible to reduce the cogging torque and torque ripple. But no certain methodological approach based on harmonic contents in the air gap field is provided for the geometries considered. In publications from [62], [63], [64], [65] and [66], the torque pulsation and the responsible harmonics based on 2D harmonic analysis are discussed. But the differences from the mentioned references in using the 2D harmonic analysis in this work is discussed in chapter 4.2.4. In this thesis a step by step approach by considering certain geometries and the influence of harmonics in air gap field for the generation of torque pulsation from these geometries are provided. Based on these analyses, a valid design rule is established for PMSM machines.

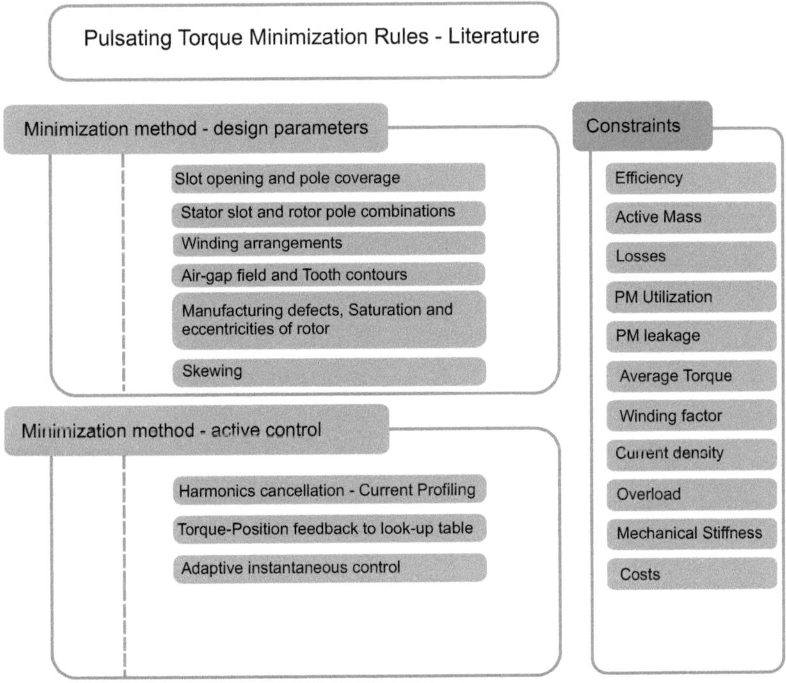

Figure 1.4.: Rules Summary

2. High Torque PMSM Application

The paper mill is selected as an application for this work. This chapter explains the drive systems in the paper mill application. The two drive systems considered are; drive with gears and direct drive. Some economic aspects for the drive system are presented. The design and working principle of PMSM for high torque direct drive application is discussed in this chapter. Finally this chapter explains the considered geometries of the work.

2.1. Paper-Mill

Paper-mill drive operates throughout the year for the production of paper. There are different kinds of paper product output such as paper for packaging, paper for newsprint etc. Due to environmental reasons, packaging paper can be produced by recycling from the garbage of papers from houses and offices. The paper-mill layout is shown in the Figure 2.1. The details in the paper-mill process is found in literature [2]. The raw material enters the mill and is rinsed with water and chemicals. The semi-solid form of garbage passes in to two major stages: wet section and drying section. The papers are pressed between the rollers that are supported by felts in the wet section to remove the water content. After the wet section paper enters the drying section where the rolling cylinders are heated with hot steam to dry the paper. The whole paper-mill room environment is of high humidity, hot and noisy.
The electrical machines in the drying section in this application are placed alternatively in pairs of drums such that two motors are coupled with torque flange. The electrical machine placed in the lower drum has different torque requirement and temperature conditions in environment than the upper drum. In recent days, the electrical machines are controlled by sophisticated drives owing to improvement in power electronics. In drying section, the conventional drive system arrangement is an electrical machine coupled to the load (drum) with gears. The

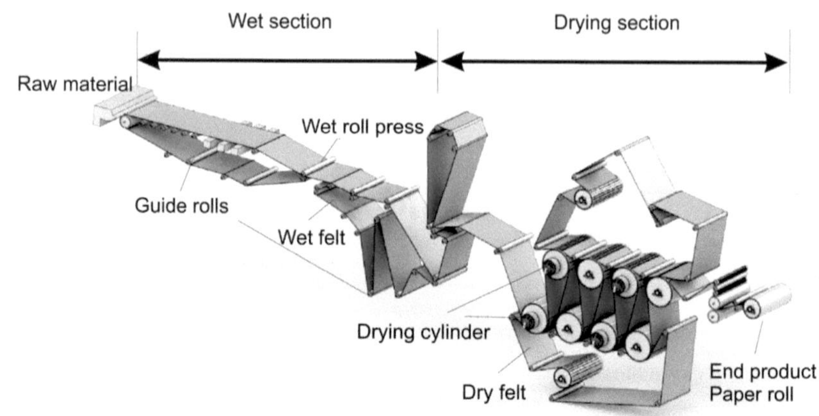

Figure 2.1.: Paper Mill.

electrical machine is usually asynchronous machine with variable speed drive. The complete drive system includes electrical machine, gear box and coupling components. Although the electrical machines have good life cycle [27], the mechanical components gears and couplings are subjected to wear and tear due to resonance [4]. In the paper-mill production, drive system component is one of the important units. Therefore, careful design of components of drive systems have benefits in terms of economy, costs, efficiency and reliability in the paper production process. In the next section the drive system which is used in drying section with and without gears has been discussed.

2.2. Drive System of Paper-mill

The electrical machine which is the main part of energy conversion can operate in all four quadrants in the torque speed curve Figure 2.2. The operating conditions for the continuous load is shown in the Figure 2.2. In this operating range the motor is within the thermal, current and voltage limits. The speed is shown inversely proportional to torque in the field weakening range. If the drive is installed with induction machine, the speed in field weakening range is not necessarily to follow this characteristics.

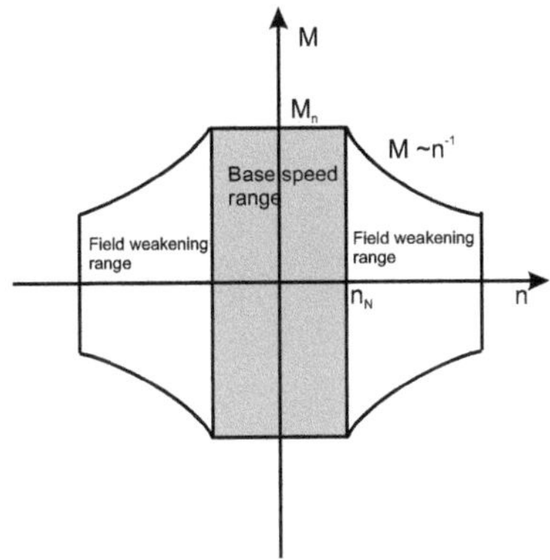

Figure 2.2.: Drive system Torque-Speed characteristics.

The speed n and the torque M are related using mechanical power as

$$P_{mech} = 2\pi M n \qquad (2.1)$$

In the paper-mill application the electrical machine operates in the motor mode (1st quadrant) and in the breaking mode (2nd quadrant). The duty cycle of the electrical motor according to the standards DIN IEC 60034-1 [28] for a typical paper mill application is shown in the Figure 2.3. It has acceleration (120 % of rated torque), constant operation (40 % to 50 % of rated torque) and deceleration (120 % of rated torque) in the load characteristics. This corresponds

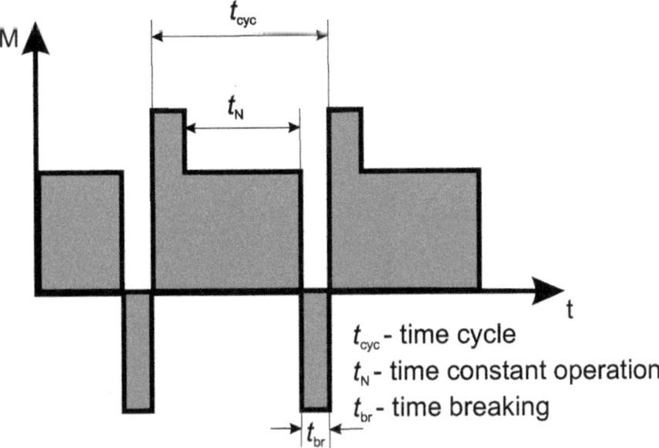

Figure 2.3.: Duty cycle [28].

to the S7 operation in the drive duty cycle and the cycle repeats periodically. The inverter is also designed for the short time overload operation. If 5 hours of maintenance in a month is considered, then the continuous annual operation of paper-mill drive is estimated as 8700 h/a.

2.2.1. Drive with Gears

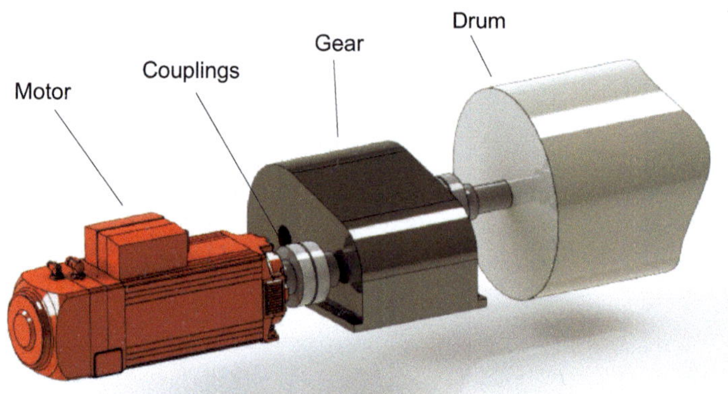

Figure 2.4.: Drive system with gear.

The basic drive configuration with gears is shown in Figure 2.4. The motor (ASM) is connected to the drum via gears and couplings. The load drum runs at low speed. ASM has disadvantage because of its low power density when it is connected as direct drive which has high torque requirement. For low speed application, according to the relation $n = f/p$, the pole-pairs of the machine should be higher. In case of ASM with increased pole-pairs, the main inductance decreases since it is inversely proportional to the number of pole-pairs. This requires more reactive power since the magnetizing current is increased. Consequently, increasing the number of pole-pairs decreases the power factor.

In paper-mill drive, the load drums are connected alternatively in pairs with more motors. All the motors should be controlled at constant speed and it requires exact information of the position. This increases complexity in control of ASM. Moreover, constant speed of ASM is difficult to control because rotor bar resistance is dependent on temperature. At high temperature the resistance changes, slip does change and time constant of the machine does also change accordingly. The torque of the machine is indirectly controlled from the stator current with information of speed and slip. This requires slip compensation in the model to meet the torque demand. Moreover, the power density or utilization of ASM is less for direct drive application. ASM can be designed for good power factor and better power density when it is connected to the load via gears.

The transmission ratio of the gears is expressed as,

$$i_g = \frac{M_{out}}{M_{in}} = \frac{r_{out}}{r_{in}} = \frac{\Omega_{min}}{\Omega_{mout}} \tag{2.2}$$

where Ω_{out} and Ω_{in} are the angular rotation, r_{out} and r_{in} are the radius of the gear and M_{out} and M_{in} are the torque. The suffix in and out is termed as input (drive side) and output (load side). The gear box serves to meet the load side requiring torque and speed. The total drive efficiency is reduced if the losses in the transmission and the couplings are considered. The drive with gears has occupied more space and has more mechanical components and reduced reliability.

2.2.2. Direct Drive

Contrary to the drive with gears, a direct drive is directly connected to the load drum. Figure 2.5 depicts a direct drive arrangement. The motor for the direct drive system is permanent

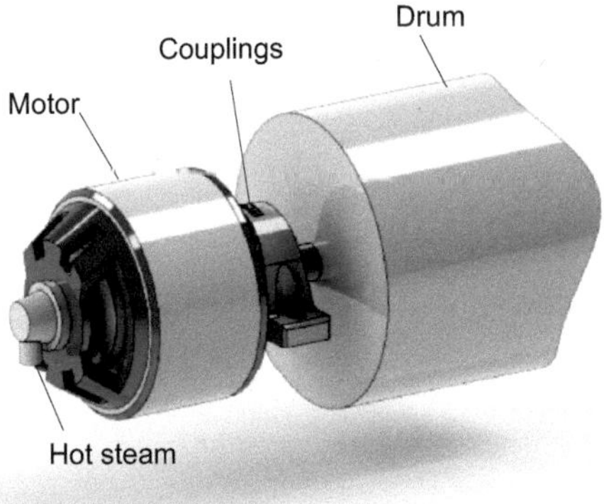

Figure 2.5.: Direct drive.

magnet synchronous motor. The hot steam inlet pipe which is used to heat the drum and outlet pipe for the condensed water are inserted through the hollow shaft of the rotor. The direct drive motor has to meet the full acceleration torque since it is directly coupled to load and at the same time the rotor magnets has to withstand high environment temperature. Moreover, the bearing of the motor has to withstand the temperature and the load. The bearing installed on the load side has to overcome the forces from the load due to huge inertia and additional vibrations. The bearing on the other side has less force when compared to the load side bearing. However, direct drive has reduced mechanical components which promises more robust and reliable design. Moreover, it occupies less space which is very important necessary during maintenance in paper-mill environment.

2.2.3. Comparison with and without Gears

A qualitative comparison of the above explained two drive system is made using the dimensions of the machine. The drive motor is designed analytically with the assumption of following basic parameters as shown in the Table2.1. Some parameters are set for the machine nominal values

Table 2.1.: Basic motor parameters.

Definition	Symbols	values
Mechanical Power	P_{mech}	70 kW
Speed	n_N	200 min^{-1}
Force density	τ_w	20/25/30 kN/m^2
Pole pitch	τ_p	26/29/33 mm
Lamination		M400-50A (Direct drive) M250-35A (with gears)
Transmission ratio	i_g	1 ... 10
No-load Flux density	$\hat{B}_{\delta_1}$	0.81 T
Current density	J_{eff}	6 A/mm^2 effective
Winding temperature		120 $°C$
Average density of active material	$\rho_{average}$	7000 kg/m^3

and other parameters are chosen empirically within the range of limits that are necessary for the rough machine design as given in books [5] and [29]. These limits are dependent on the applications. For example, in automotive application the machine air gap cannot be selected as given in the formulae in books [5] and [29]. Usual way to begin the electrical machine design approach is from the force density.

$$\tau_w = \frac{\hat{A}_{s_1} \hat{B}_{\delta_1}}{2} \cos(\alpha) \tag{2.3}$$

where $\hat{A}_{s1}$ is the current sheet, $\hat{B}_{\delta 1}$ is the no-load flux density and α is the angle between both the fields. For this simple analytical approach the main dimensions of the machine bore diameter D and length L can be related with the expression.

$$D = \sqrt[3]{\frac{2\frac{D}{L}\frac{M_a}{i_g}}{\pi \tau_w}} \tag{2.4}$$

where M_a is the nominal torque and i_g is the transmission ratio. Similar to the dimensions (D and L), for the considered power, the mass of the active region are compared. From the dimensions D and L, the volume of the copper with end winding and iron are determined with some factors which are necessary to estimate the mass of the active parts. The total mass including copper and iron is related with the expression,

$$m_{active} = \rho_{average}(3h_j + 0.5h_n)\pi DL + m_{cu} \tag{2.5}$$

where $\rho_{average}$ is the average material density, h_j is the height of the yoke, h_n is the height of the teeth and m_{cu} is the mass of the copper and it is defined as.

$$m_{cu} = \frac{\rho_{cu}\sigma_{cu}}{J_{eff}^2}P_{v,cu} \tag{2.6}$$

where ρ_{cu} and σ_{cu} are the copper material density and conductivity respectively, J_{eff} is the current density and $P_{v,cu}$ is the loss of the copper. The first part of the right hand side (RHS) of the Equation 2.5 factor includes the consideration of yoke, teeth and magnet for PMSM or copper for ASM. The length of the end winding is related with factor c_1. However, the factor is adapted according to the winding type, distributed or fractional slot winding arrangement. An annular lamination core of fixed cylinder(Stator) and rotating cylinder(Rotor) separated by an air gap of a magnetic circuit is shown in the Figure 2.6. The effect of slot is neglected.

Based on this approach two drive systems (with and without gear) are considered. As an example, the calculation is made for a mechanical power of 70 kW and a speed of 200 min^{-1}. Figures 2.7 and 2.8 show the dimension and mass of direct drive motor. The calculations are carried out for three different force density for this application and the pole pitch is varied

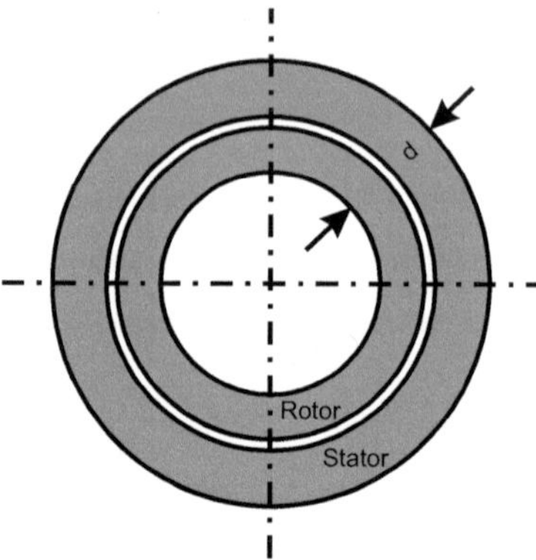

Figure 2.6.: Cylindrical magnetic circuit.

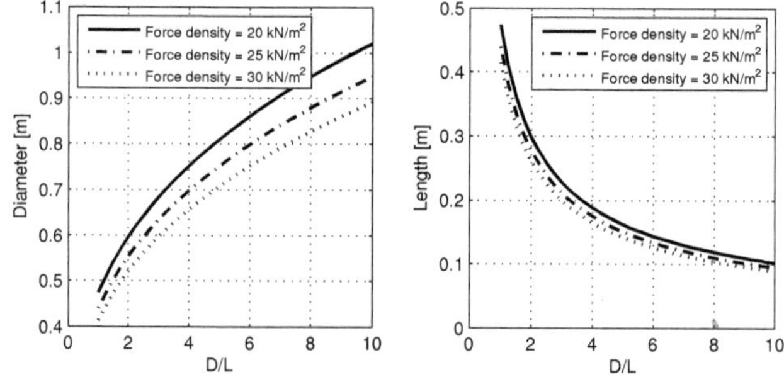

Figure 2.7.: Variation of diameter and length for D/L.

between 26 mm and 33 mm. Figures 2.7 and 2.8 also show the tendency of diameter, length and mass for D/L factor. For the mass, the force density of 30 kN/m^2 is taken as constant and the pole pitch is varied. Figures 2.9 and 2.10 provide details about the motor with gears.

The curves in Figures 2.9 and 2.10 show the dimensions and mass for different gear ratio by varying the parameters such as D/L and pole pitch. The speed is calculated by the product of gear ratio with the base speed. For the mass calculation, the force density of 30 kN/m^2 and the D/L ratio of two are taken as constant. When compared with the direct drive, the mass of the motor for drive with gears is less for higher gear ratio. In the paper [30], a cost based comparison with and without gears has been made for the automotive application. In this publication, the motor with gear has been concluded as the best solution. However,

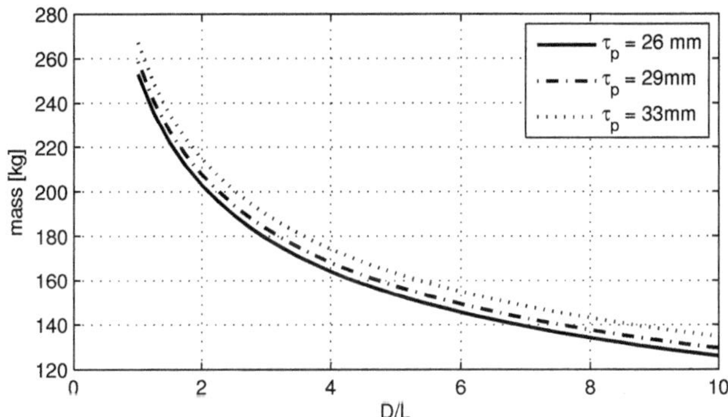

Figure 2.8.: Active mass for different D/L $Forcedensity = 30kN/m^2$.

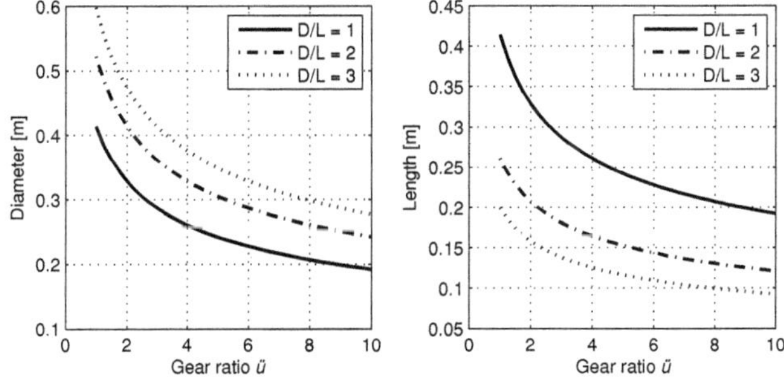

Figure 2.9.: Variation of diameter and length for i_g $Forcedensity - 30kN/m^2$

concerning the total weight including gear and the volume requirements, the direct drive is the best solution. In contrary to the automotive application, the direct drive application other factors such as mechanical problems, space, dynamics, costs, half load efficiency, productivity and economical aspects should be taken in to account for the overall drive comparison. For example, a 6 MW direct drive for wind generator [4] has more mass than the generator with gear. Apart from dimensions and mass of active material, the overall drive comparison based on economic aspects are explained in the following section.

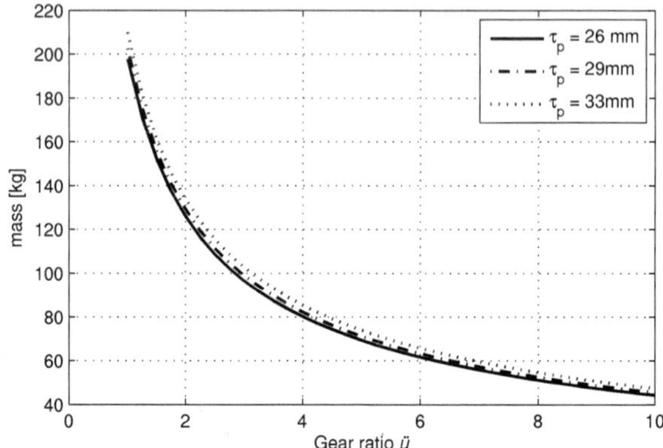

Figure 2.10.: Active mass for different i_g $Forcedensity = 30kN/m^2$, $D/L = 2$.

2.2.4. Economical and Ecological Aspects

Since there is a demand for electrical energy, a part of energy can be saved and CO_2 emission can be controlled if an industry such as paper-mill are replaced with new technologies. A calculation is derived to determine the economical aspects of the direct drive. Consider that there are 4 motors installed in the drying section (2.1). The mechanical power of 70 kW is assumed. Assuming 5 hours of maintenance work in the paper-mill in a month, then the annual running period of the drive is 8700 h. The price of energy for typical industry application is $K_H = 0.1(EUR/kW)$. The CO_2 emission is estimated by assuming that 1 kWh of energy consumption emits 0.2 kg of CO_2. The estimated price and the CO_2 emission are shown in Table 2.2.   Thus in a direct drive system, a cost of 11664 EUR is saved in a year when compared to the drive with gears for one drying section with 4 motors installed . At the same

Table 2.2.: Economic Calculations.

Definition	Direct drive	Drive with gear
Efficiency [%]	93	89
Electrical Power [kW]	75.3	78.7
Energy consumption per year [kWh/a]	655110	684269.7
Cost of Energy consumption [EUR]	65511	68427
CO_2 Emission [kg]	131022	136853.9

time, less amount of CO_2 is exposed to the environment which in Europe requires less CO_2 certificates.

2.3. Application of PMSM in Paper-Mill

Focusing the attention towards PMSM as an alternative direct drive solution for the paper-mill application, PMSM shows excellent power density and efficiency [31]. In publication [32] different motor types for 75 kW power are compared and it is shown that PMSM has high power density. Permanent magnet machine with fractional slot winding arrangement features advantageous properties such as

- High power density.

- Reduced copper loss and increased efficiency due to the short winding overhang.

- Good efficiency at partial load, as an example half load.

- Economical and environmental benefits.

The disadvantages in the direct drive motors are costs and availability of magnets. However, these could be optimized during the design process. Direct drive would improve the life cycle characteristic of the drive system. Based on the discussion in the previous sections 2.2 and 2.3, a general comparison is summarized in Table 2.3.

Table 2.3.: Drive Comparison.

Definition	Direct drive	Drive with gear
Efficiency	++	+
Power density	++	-
Costs	+	++
Control technology	++	+
Space	++	–
Mass	++	-
Mechanical problems / Reliability	+	–
Economy / Ecology	++	+

2.4. Generic Design and Working Principle of PMSM

In this dissertation work direct drive having PMSM with fractional slot winding arrangement
is studied. The additional design specifications that are necessary for the generic design of
the motor apart from the Table 2.1 is tabulated in Table 2.4. A general electrical machine
design approach is explained in the literature [6], [29], and [5]. Figure 2.11 shows an overview

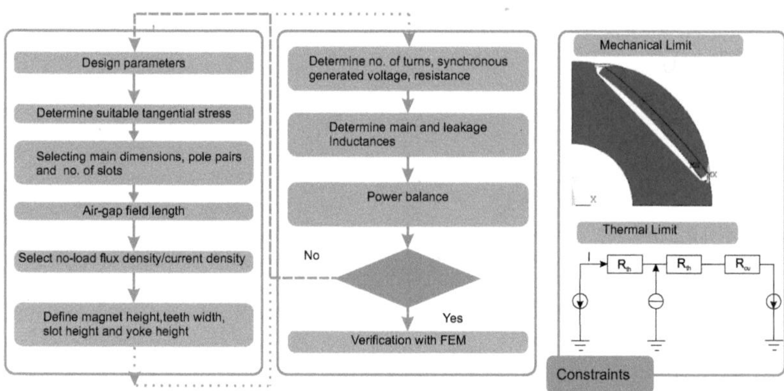

Figure 2.11.: Electrical machine design approach.

of methods involved in dimensioning the electrical machine. From the force density, the active
length of the machine is determined using the relation,

$$L_{fe} = \frac{M}{2\pi r_{st}\tau_w}$$

(2.7)

Table 2.4.: Design specification parameters.

Definition	values	Units
Outer Diameter	up to 650	mm
Air gap length	1.5...2	mm
Tooth flux density	1.4...2.1	T
Yoke flux density	0.8...1.5	T
Current density	6	A/mm^2
Cooling type	Water Cooled	
Enviromental temperature	80	$^\circ C$

where M is the torque and r_{st} is the bore radius. The equation involved in determining the other parameters is not shown here. The equation that applies to determine no-load flux density and calculation of magnet parameters will be discussed in the next chapter. This analytical design calculation can be executed iteratively until required power balance is determined. Once the machine dimensions are obtained the next step is to make precise calculation using FEM program. Apart from the electromagnetic limitation, it is to be noted that the machine parameters must withstand mechanical and thermal limits. These are mentioned as constraints in Figure 2.11. When the mechanical and thermal limitations are not satisfied then parameters such as current density and magnetic circuit dimension are iterated until best machine is obtained. In this design process, noise and vibrations, costs, bearing design, shaft design etc. are not considered. The analytical iterative process is also made with the help of optimization routine (Chapter 5). However, better insight of the electrical machine is important to decide initial parameters, objective function and constraints to select the best machine out of it.

2.4.1. Physical Description of PMSM

PMSM which works on the principle of classical AC electrical excited synchronous machine (SM) generates steady state torque when stator field and rotor field rotate at same speed relative to each other. The stator of the machine is made up of three phase winding arrangement similar to the SM or ASM. Instead of DC rotor excitation, the rotor field is excited by permanent magnets. There are various geometries of stator and rotor ([29] and [33]) for PMSM.

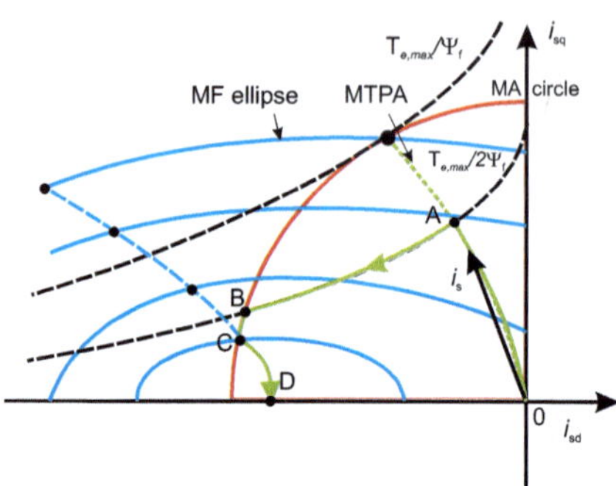

Figure 2.12.: Field weakening - classical method [34].

The arrangement of stator and rotor considered in this work is described in the next section. The phasor representation and the steady state voltage equations provide performance of the machine for the various operating points. In steady state operation, the flux linkage is constant and thus $d\psi/dt = 0$.

Including the resistance the steady state voltage equation in motor mode is

$$U_d = R_s I_d + j X_q I_q \tag{2.8}$$

$$U_q = R_s I_q + j X_d I_d + j \omega_m \psi_{PM} \tag{2.9}$$

where ω_m is the angular speed, $X_d = \omega_m L_d$ the d-axis reactance and $X_q = \omega_m L_q$ is the q-axis reactance. The last term in equation 2.9 is synchronous generated voltage. The electro-magnetic torque together with the reluctance torque is given by.

$$M = \frac{3p}{2}(\psi_{PM} I_q + (L_d - L_q) I_d I_q) \tag{2.10}$$

where L_d and L_q are the stator d-axis and q-axis inductances.

The machine is fed with frequency inverters with variable frequency and voltage. The operating points and control of PMSM for different speeds are explained in book [34]. The current components in abscissa(d) and ordinates(q) of PMSM with unequal inductances L_d and L_q for the motor operation are shown in Figure 2.12. In Figure 2.12 MTPA denotes maximum torque per ampere, MF ellipse (blue) represents the voltage limit and MA circle (red) denotes the current limit. The current trajectory is increased along 0A in MTPA line until it reaches its thermal limits. In Figure 2.12, the current vector is projected for half of the rated point. Between line AB, the torque can be kept constant following the hyperbolic path (black dashed-line) under MF ellipse. This region is called constant torque region and the speed is increased by varying the maximum flux linkage. Between lines B and C, the speed is increased by field weakening and following MA circle where current reaches its limits. Further increase in speed is possible following the line CD. The necessary current and voltage equations for the operating points mentioned are found in [34].

The method shown in Figure 2.12 discusses about the operational limits of the motor. The other method to control PMSM is to introduce an intermediate region by removing the winding turns to increase the speed. The field weakening begins before the voltage reaches its limits. This approach is shown in Figure 2.13. Thus in the intermediate region the speed is increased by increasing the voltage with phase change in the voltage slope. At the same time, the inverter fed current is reduced. Now, the voltage can be increased until its rated limit. Beyond the intermediate region the speed can be increased by weakening the flux. The angle beta (green

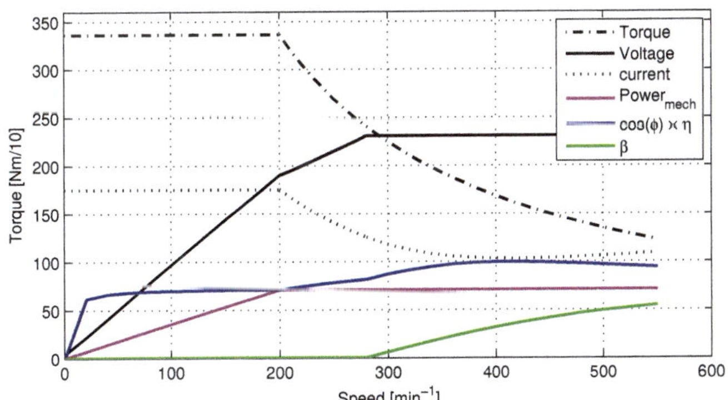

Figure 2.13.: Field weakening - intermediate region.

curve) in Figure 2.13 indicates the starting of the field weakening region. The advantage of this method is that, at higher speeds there is a reserve for current until the inverter fed supply current reaches its limit, i.e the current limit is reached later at higher speed. This strategy has influenced the loss characteristics because of higher stator currents as studied in the Master thesis work [35].

2.4.2. Selection of Ferromagnetic Materials

The magnetic materials are classified in to three kinds based on the susceptibility: diamagnetic, paramagnetic and ferromagnetic. The ferromagnetic materials are magnetized externally and classified further in to hard and soft magnetic materials. In the hard magnetic materials substance, the magnetization persists when the external field is removed and those materials are composed of compound of alloy. These materials are called Permanent Magnets. For example, Alnico is an alloy of Iron-Aluminium-Nickel-Cobalt, SmCo is an alloy of Samarium-Cobalt and NdFeB is an alloy of Neodymium-Iron-Boron. The demagnetization curve in the

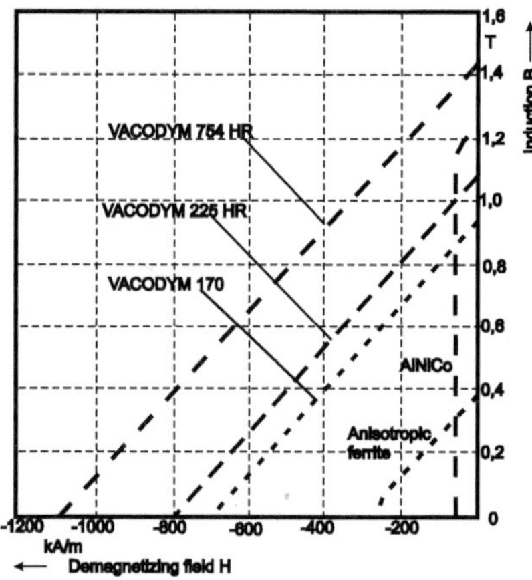

Figure 2.14.: Permanent magnet material characteristics [36].

hysteresis curve allows to select suitable permanent magnet material. Such material should posses high remenance and coercivity. In the application like paper-mill, the environmental temperature is higher and the utilization of the permanent magnet should be greater. At the same time, the magnet should not be demagnetized when the machine operates under load and also in field weakening. Based on the product catalogue (Figure 2.14), NdFeB type permanent magnet is selected for this work.

2.5. Considered Geometry of PMSM Stator/Rotor

The developed view of considered stator geometry in this work is depicted in Figures 2.15 and 2.16 for one elementary machine. The conventions used to represent an elementary machine in Figures 2.15 and 2.16 are from references [37] and [38], it is divided in to zones, coil groups and repetition factors and is represented as $< z_{sp} - r_1 - r_2 - z_{pm} >$. The $< 12 - 2 - 2 - 10 >$

Figure 2.15.: Elementary machine winding arrangement $< 3 - 1 - 1 - 2 >$.

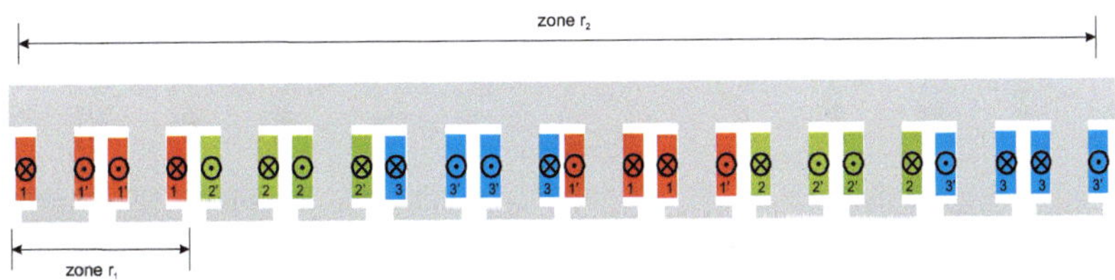

Figure 2.16.: Elementary machine winding arrangement $< 12 - 2 - 2 - 10 >$.

arrangement is shown in Figure 2.16 which means, it has 12 coils ($Z_{sp} = 3r_1r_2$) in a zone of repetitive factor $r_1 = 2$ and repeated $r_2 = 2$ coil groups in the elementary machine with 10 permanent magnets. The repetition factor $r_2 = 2$ forms coil groups with all three phases. The other stator arrangement considered is $< 3 - 1 - 1 - 2 >$ (Figure 2.15). Both these arrangements are fractional-slot or tooth-coil type of two layer winding with number of slots per pole per phase less than one ($q < 1$).

A systematic approach for fractional slot winding arrangement having $q < 1$ for different combinations are studied in details in the work from [38]. This type of arrangement has higher harmonics and sub-harmonics (Figure 3.20 b) that are responsible for losses in the conducting region of rotor. The sub-harmonics can penetrate deep in to the rotating part and increase the losses in the rotor [39]. Of course, harmonic waves can also create torque pulsations and those harmonics that are responsible for torque pulsation are the focus of this work. There are several ways to eliminate the harmonic content by modifying the magnetomotiveforce MMF pattern. Those methods are proven in reference [14]. By varying the number of turns within the coil, the winding factor of the harmonics is effectively reduced. Thus the magnitude of the harmonics is eliminated. As discussed in paper [40], the number of turns is different for the coils of arrangement $< 12 - 2 - 2 - 10 >$. By selecting such a multi-layer arrangement as discussed in the paper [40], the winding factor of the sub harmonic component is eliminated which is responsible for eddy current losses in the permanent magnets. Instead of reducing the

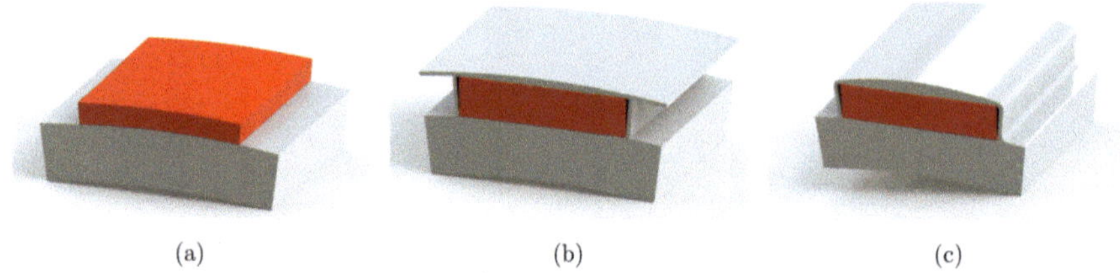

(a)   (b)   (c)

Figure 2.17.: Rotor configurations. (a) Surface mounted (b) Shell-type (c) Pole-cap.

magnet losses there is increase in the winding copper losses. By varying the stator teeth, 7th harmonics can be eliminated [14].

The winding factors of fractional-slot winding is also discussed in [38]. The winding factors for the working harmonics of the two winding arrangement type that are used in this work are 0.866 for $< 3 - 1 - 1 - 2 >$ and 0.933 for $< 12 - 2 - 2 - 10 >$.

Figure 2.17 illustrates the three different rotor topologies that are analyzed in this work. The first rotor type is classical surface mounted machine. The next two types are rectangular protected magnets, one with shell type laminations and the other with pole-cap laminations. The advantage of protected magnet arrangement is that, it is easy to assemble and the manufacturing cost is less. The magnet considered in this work is capable of working at temperature up to 120°C. For the paper mill application the inner surface of the rotor is covered by special thermal insulation with glass-fibre-reinforce-plastic material which prevents the heat from the steam armature to creep in to the rotor. Thus the magnet is prevented from heat and is not cooled directly. Also, due to the mechanical protection of the permanent magnet with iron ring and due to insulation of the inner surface of rotor, the utilization of the magnet will be higher.

Table 2.5.: Geometries combinations.

Definition	Stator arrangement	Rotor type
D1	$< 3 - 1 - 1 - 2 >$	Surface mounted
D2	$< 3 - 1 - 1 - 2 >$	Shell-type
D3	$< 3 - 1 - 1 - 2 >$	Pole-cap
D4	$< 12 - 2 - 2 - 10 >$	Surface mounted
D5	$< 12 - 2 - 2 - 10 >$	Shell-type
D6	$< 12 - 2 - 2 - 10 >$	Pole-cap

The combinations from stator and rotor arrangements concludes with six different forms. The chosen Pseudonym of those six combinations is summarized in Table 2.5.

3. General Derivation of Force in PMSM

The sources for the torque pulsations are harmonics from the stator and rotor fields and these fields are distributed in space and time. In this chapter, certain mathematical rules to derive the torque pulsation has been discussed. It is wise to understand the spatial harmonic distributions in the fields of rotor and stator before deriving the force in electro-magnetic energy conversion. These harmonics are further superimposed with various permeances in the magnetic circuit that may increase or generate additional harmonic contents in the air gap field solutions. These field solutions are derived analytically and compared with FEM in this chapter.

3.1. Classification of Parasitic Torque

The occurrence of parasitic torque in the PMSM is classified from the origin of harmonic sources. In PMSM, the parasitic torque is composed of electromagnetic torque ripple and cogging torque. Cogging torque results from harmonics present in the no-load flux density wave distribution of permanent magnets. In this case, the harmonic sources are no-load flux density wave distribution and superposition of slot openings on the no-load flux density. The slot opening that modulate the air gap flux density is represented as slot permeance function. These permeance variations are influenced if the slot opening height is very small because it can saturate the tooth tips even under no-load conditions and there is fictitious widening of the slot opening.

Under load, torque oscillation is also caused by the interaction of magnet harmonics with the space harmonics generated by the spatial distribution of the current sheet. Here the slotting has further consequences by superimposing on the air gap flux density. All the permeance functions which are shown in Figure 3.1 interact with stator and rotor fields and are responsible for the generation of parasitic torque. In this work, it is necessary to discuss the process of generation

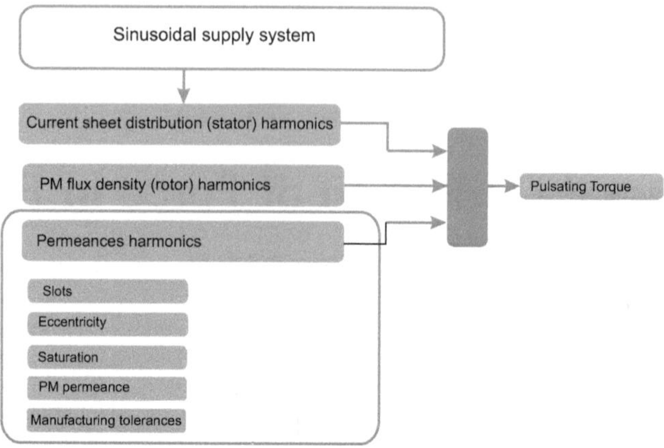

Figure 3.1.: Sources of Harmonics.

of harmonics in the air gap flux density and it is outlined in the Figure 3.1. Mathematically, flux density distribution from stator, rotor and permeance variations are expressed using Fourier series. In the current sheet waves and the no-load flux density waves, the permeance functions are superimposed. As discussed in Chapter 1, the source of harmonics due to inverter fed supply system is neglected and thus the current is considered as pure sinusoidal wave for the calculation. In the following section, the permeance functions and flux density wave distribution in the air gap due to stator and rotor and generation of torque are discussed.

3.2. Components of Air Gap Flux Density and Permeances

3.2.1. No-load Flux Density

In this section, the two dimensional field equation in polar coordinates is solved to determine the flux density due to permanent magnet in the air gap. For the field calculation, the analytical method used in [41] is derived and is used in this work. Following assumptions are considered in solving the Possionian or Laplacian equation analytically.

- The machine geometry is assumed as two regions cylindrical model as shown in the Figure 3.2.

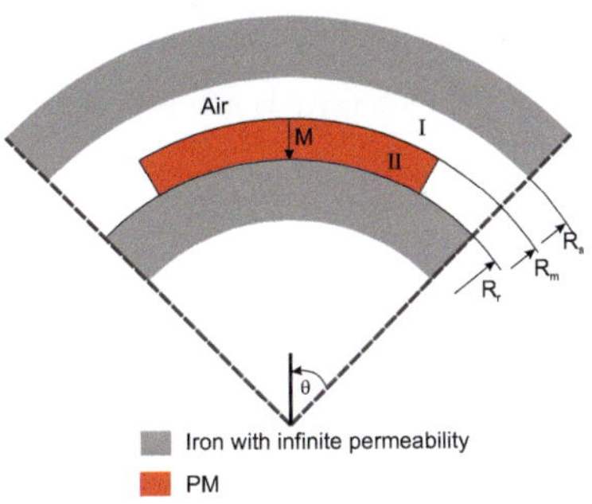

Figure 3.2.: Geometry representation.

- The relative permeability of stator and rotor magnetic material assumed to be infinitely permeable, thus the field is perpendicular to the iron.

- Since it is a two dimensional solution, end effects in the axial direction are neglected.

- For the permanent magnets the material is linear isotropic. Only the radial magnetization is considered and it is polarized alternatively (north pole and south pole).

The field solution is a two region problem. As shown in Figure 3.2, the bore radius is at R_s, radius until magnet is R_m and the inner boundary radius is R_r. The magnetic flux density $\vec{B}$ and the magnetic field intensity $\vec{H}$ are expressed as.

for the airgap,

$$\vec{B}_I = \mu_0 \vec{H}_I \tag{3.1}$$

for the permanent magnet,

$$\vec{B}_{II} = \mu_M \vec{H}_{II} + \mu_0 \vec{M} \tag{3.2}$$

where $\mu_M = \mu_0 \mu_{rM}$ is product of permeability in free space and relative permeability and $\vec{M}$ is the magnetization of the permanent magnet.

The magnetostatic field equation in differential form for current free region is written as,

$$\nabla \vec{B} = 0 \tag{3.3}$$

$$\nabla \times \vec{B} = 0 \tag{3.4}$$

If a magnetic scalar potential φ_M is introduced as work done in moving an unit pole from infinity to a point, then magnetic field intensity is gradient of a scalar potential φ_M:

$$\vec{H} = -\nabla\varphi_M \qquad (3.5)$$

According to Gauss law, the net flux crossing is zero and hence in free space after making substitution of Equation 3.5 in Equation 3.3

$$\mu_0\nabla(-\nabla\varphi_M) = 0 \qquad (3.6)$$

This satisfies the Laplace equation in the air gap region I and it is given by,

$$\nabla^2(\varphi_I) = 0 \qquad (3.7)$$

In the permanent magnet region, the volume charge model is used [42]. The volume charge density is given by

$$\rho_M = -\nabla \cdot \vec{M} \qquad (3.8)$$

Now the gradient of the magnetic field is expressed as,

$$\nabla\vec{B} = \rho_M \qquad (3.9)$$

On substituting the equations 3.5 and 3.8 in the previous equation 3.9, the Poisson's equation representing the magnetization for the magnet region is given as,

$$\nabla^2(\varphi_{II}) = \frac{\nabla \cdot \vec{M}}{\mu_r} \qquad (3.10)$$

Now it is necessary to define the magnetization distribution. The magnetization of permanent magnet is decomposed in to two components in polar coordinates and it is expressed as,

$$\vec{M} = M_r\vec{i_r} + M_\theta\vec{i_\theta} \qquad (3.11)$$

Only radial magnetization M_r is considered for the field calculation and thus the tangential component M_θ does not exist. If the magnetization function is considered as rectangular distribution as shown in Figure 3.3, the corresponding function can be expressed

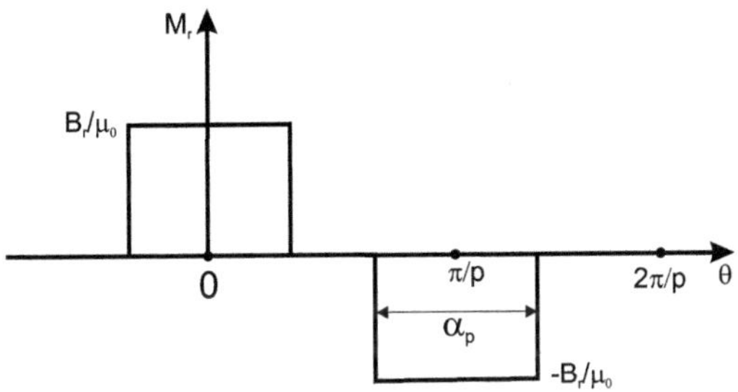

Figure 3.3.: PM field distribution.

$$f(\theta) = \begin{cases} B_r/\mu_0 & \text{for } -\alpha_p/2p < x < \alpha_p/2p \\ -B_r/\mu_0 & \text{for } -\alpha_p\pi/2p < x < \alpha_p\pi/2p \\ 0 & \text{otherwise} \end{cases}$$

The Fourier series of the above function is given in Equations 3.12 and 3.13 ([41],[48]).

$$M_r = \sum_{n=1,3,5...}^{\infty} M_{rn}cos(np\theta) \tag{3.12}$$

$$M_\theta = \sum_{n=1,3,5...}^{\infty} M_{\theta n} \sin(np\theta) \tag{3.13}$$

where θ is the angular position and p is the number of pole pairs. However for the radial magnetization, the Equation 3.13 is zero. Finally the given odd function shown in the Figure is given by,

$$M_{rn} = \frac{B_r}{\mu_0} \frac{4}{n\pi} \sin \frac{n\pi\alpha_p}{2} \tag{3.14}$$

where α_p is the pole coverage. Let r represent the radial outward direction and θ represents the peripheral couterclock wise direction, the divergence of the magnetization is,

$$\nabla \cdot \vec{M} = M_r + r\frac{\partial M_r}{\partial r} + \frac{\partial M_\theta}{\partial \theta} = \frac{1}{r} \sum_{n=1,3,5...}^{\infty} M_{rn}cos(np\theta) \tag{3.15}$$

The scalar magnetic potential (Equation 3.6) is considered for the Laplacian field for the air gap region(Figure 3.2). The corresponding Laplacian equation in polar coordinates (r, θ) is

$$r^2\frac{\partial^2 \varphi_I}{\partial r^2} + r\frac{\partial \varphi_I}{\partial r} + r^2\frac{\partial^2 \varphi_I}{\partial \theta^2} = 0 \tag{3.16}$$

and Poissonian fields (Equation3.10) for the magnet region in polar coordinates is expressed as,

$$r^2\frac{\partial^2\varphi_I}{\partial r^2} + r\frac{\partial\varphi_I}{\partial r} + r^2\frac{\partial^2\varphi_I}{\partial\theta^2} = \frac{1}{r}\sum_{n=1,3,5...}^{\infty} M_{rn}cos(np\theta) \tag{3.17}$$

3.2.2. Solution of the PDE (Method of Separation of Variables)

The homogeneous linear partial differential equation(PDE) is solved by the method of separation of variables. The complete solution has two parts,

1. Complementary Function (CF)

2. Particular Integral (PI)

Let the solution of equation(3.16) be of the form

$$\varphi_I(r,\theta) = R(r) \cdot \varphi(\theta) \tag{3.18}$$

where R is a function of r only and φ is a function of θ only.

Substituting (3.18) in (3.16), the general solution satisfying the Laplace equation is in the form,

$$\varphi_{CF}(r,\theta) = (Ar^{np} + Br^{-np})(C\cos(np\theta) + D\sin(np\theta)) \tag{3.19}$$

where constants A,B,C and D are to be determined using boundary conditions. This solution (Equation 3.19) holds for the air gap region. For the magnet region, a particular function is added to get the complete solution. It is represented as

$$\varphi_{PI}(r,\theta) = \sum_{n=1,3,5...}^{\infty} \frac{M_n}{\mu_r[1-(np)^2]}r\cos(np\theta) \tag{3.20}$$

Then it is possible to write the total solution for each region as:

for air gap region $(R_s > r > R_m)$,

$$\varphi_I(r,\theta) = (A_1 r^{np} + B_1 r^{-np})cos(np\theta) \tag{3.21}$$

for PM region $(R_m > r > R_r)$,

$$\varphi_{II}(r,\theta) = (A_2 r^{np} + B_2 r^{-np})cos(np\theta) + \frac{M_n}{\mu_r[1-(np)^2]}r\cos(np\theta) \tag{3.22}$$

The boundary conditions at the inner and outer boundaries are assumed so that the tangential field vanishes. From this condition the constants B_1 and B_2 are determined.

$$B_1 = -A_1 R_s^{2np} \tag{3.23}$$

$$B_2 = -A_2 R_r^{2np} - \frac{M_n}{\mu_r[1 - (np)^2]} \cdot R_r^{np+1} \tag{3.24}$$

The next boundary conditions that satisfy at the interface of two regions are tangential component H_θ and radial component B_r which are continous. These are given by,

$$H_\theta = -\frac{1}{r}\frac{\partial \varphi}{\partial \theta} \tag{3.25}$$

$$B_r = \mu_0 \left(-\frac{1}{r}\frac{\partial \varphi}{\partial r} + M_r \right) \tag{3.26}$$

Evaluating the boundary conditions in Equations (3.25 and 3.26) and solving the linear equations gives the coefficients A_1 and A_2. Thus at the air gap ($r = R_m$) the solution of flux density in the radial direction is,

$$B_{rI} = -\mu_0(np)A_1[r^{np-1} + R_s^{2np}r^{-np-1}]\cos(np\theta) \tag{3.27}$$

where the coefficient A_1 is expressed as,

$$A_1 = \frac{M_n}{[1 \quad (np)^2]\mu_,}R_m^{-(np-1)}\frac{[R_m^{2np}(np - 1) + 2R_r^{np+1}R_m^{np-1} - R_r^{2np}(np + 1)]}{\frac{\mu_r+1}{\mu_r}[R_s^{2np} - R_r^{2np}] - \frac{\mu_r-1}{\mu_,}[R_m^{2np} - R_s^{2np}R_r^{2np}R_m^{-2np}]} \tag{3.28}$$

Similarly the solution for the flux density in the tangential direction is,

$$B_{\theta I} = \mu_0(np)A_1[r^{np-1} - R_s^{2np}r^{-np-1}]\sin(np\theta) \tag{3.29}$$

This solution is valid for $np \neq 1$. The special case $np = 1$ is provided in reference [41]. The field solution for the air gap based on vector potential is also used in this work. The derivation of this analytical solution is provided in the lecture script [43]. The parameters that are going to be used for no-load flux density distribution are tabulated (Table 3.1).

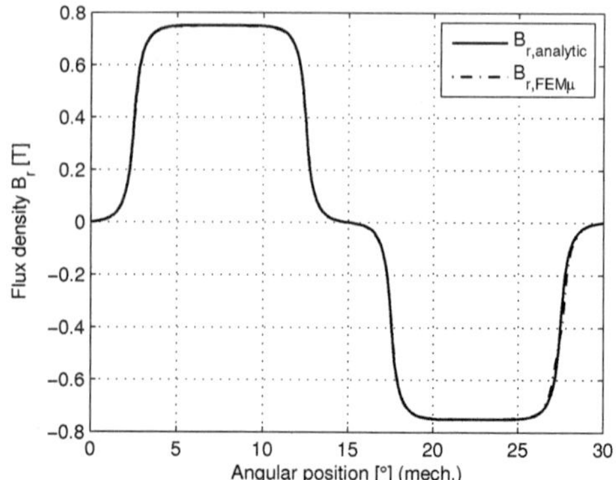

Figure 3.4.: No-load radial flux density (D1) at middle of the air gap.

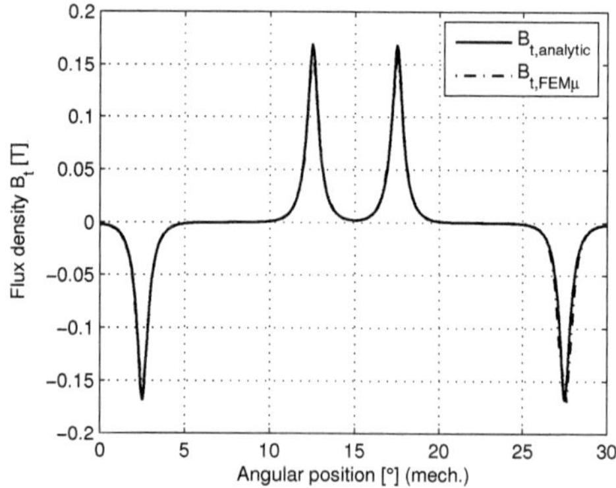

Figure 3.5.: No-load tangential flux density (D1) at middle of the air gap.

Table 3.1.: Design parameters

Definition	values	Units
Polepairs(D1/D4)	12/15	
Slots(D1/D4)	36	
Bore Diameter	500	mm
Air gap length	2	mm
Magnet height	5.5	mm
Pole coverage	0.67	
Remanence flux density	1.05	T
Relative permeability PM	1.051	

In figures 3.4 and 3.5, the no-load flux densities of radial and tangential components of the geometry D1 for two pole periods $(2 \cdot \tau_p)$ are shown. The flux density calculated analytically is compared with FEM results without slots at the middle of the air gap. In the FEM calculation, linear magnetic material of permeability $\mu_r = 5000$ is considered. Analytically calculated results match exactly the FEM results. In general the air gap flux density is expressed as,

$$B_\delta(x_r, t) = \sum_{\mu=1,3,5...}^{\infty} B_\mu cos(\nu\pi/\tau_p - \omega t - \varphi) \tag{3.30}$$

where μ is the odd harmonics of order $\mu = p, 3p, 5p, ...$

3.2.3. Magnet Leakage/Magnet Utilization

The no-load flux density depends on the height of permanent magnets and mechanical air gap. The considered geometry types for magnet leakage calculations are D1 and D2. The characteristics of no-load flux density (peak value of fundamental) against permanent magnet (PM) height for different air gap length are shown in Figure 3.6. The ratio of pole-arc to pole-pitch is approximately 0.67. The no-load flux density is calculated with the help of FEM program for different variations of height of magnet and mechanical air gap. The variation of the air gap is from $0.5mm$ to $10mm$ but the variation of magnet height is different for two geometries. Figure 3.6 shows that no-load flux density increases for certain heights and then is constant. This is due to the leakage of the permanent magnets. It should be noted that to have a flux density of $0.8\,T$ in the air gap for $2mm$ of air gap length, the magnet height is approximately doubled for geometry D2 when compared with D1.

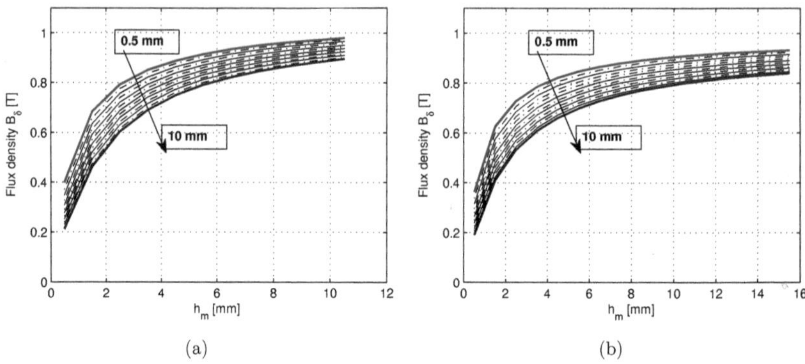

Figure 3.6.: Rotor configurations. (a) Surface mounted D1 (b) Shell-type D2.

The utilization of the magnet which is torque per magnet mass is calculated for all geometries (D1 to D6). Figure 3.7 shows the torque for different magnet masses for all geometries. The masses are calculated by keeping the pole coverage constant for given active length of the machine by varying the height of the magnet. The air gap length is kept at $2mm$ and the rated current is imposed in the stator. The surface mounted topology requires less magnet material when compared to the shell-type or pole-cap structure. The leakage is more in shell-type and pole-cap type and it requires more magnet material to achieve the same torque as surface mounted PMSM. For example, in order to have $3\,kNM$ torque the utilization of magnet for D1 is 11.3 % better than compared with D2. The utilization is also reduced for the geometries D4/D5/D6 which have increased pole-pairs. In all the cases the decrease of PM utilization is due to leakage and it also decreases due to reduction in the fundamental flux density in the air gap and consequently the torque is reduced. Although the magnet volume is increased, there is no increase in torque in the machine beyond certain height of magnets.

3.2.4. Armature Current Sheet

Each machine type has its own current sheet distribution. The current sheet is a thin strip of conducting layer attached to the stator. The current sheet is studied for three phase system

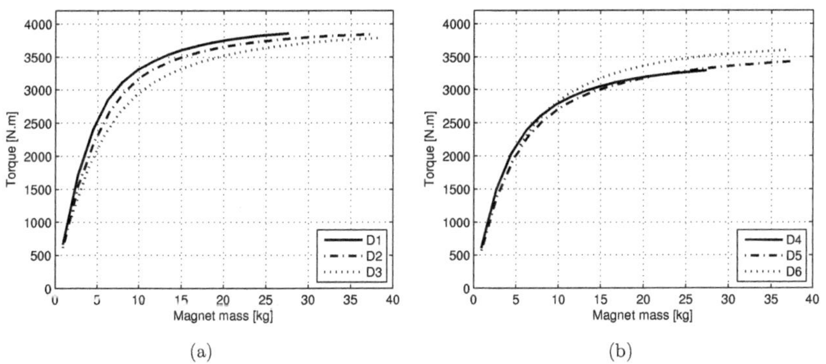

Figure 3.7.: Magnet Masses. (a) Geometries D1/D2/D3 (b) Geometries D4/D5/D6.

of fractional-slot winding arrangement with number of slots per pole per phase less than one. The wave representation of current sheet is written as,

$$A_s(x_s, t) = \sum_{\nu=1,3,5...}^{\infty} A_\nu cos(\nu\pi/\tau_p - \omega t)$$ (3.31)

where ν is the winding harmonics that has been discussed again later in this chapter. A_ν is expressed as,

$$A_\nu = \sqrt{2}\xi_{w\nu} \cdot \frac{m w_s I_s}{p\tau_p}$$ (3.32)

where $\xi_{w\nu}$ is the winding factor, m is the number of phases, w_s is the number of turns and I_s is the effective current. Figures 3.8 and 3.9 show the current sheet distribution and harmonics of the considered stator type.

The developed Figures 3.8 and 3.9 shown are made for one elementary machine. Harmonics are assumed to be produced by sets of additional poles superimposed on the fundamental poles. The fractional slot type $< 12-2-2-10 >$ uses 5th ordinal for the generation of the torque and it also contains sub-harmonics. Thus for this winding arrangement, the fundamental component 1st ordinal is not the working harmonic but the 5th ordinal is the working harmonic. The other harmonics of current sheet may also generate torque if they find a corresponding wave of the field distribution with the same pole pitch. If the machine is selected with torque producing harmonics of higher order the required frequency will also be higher for the same speed leading

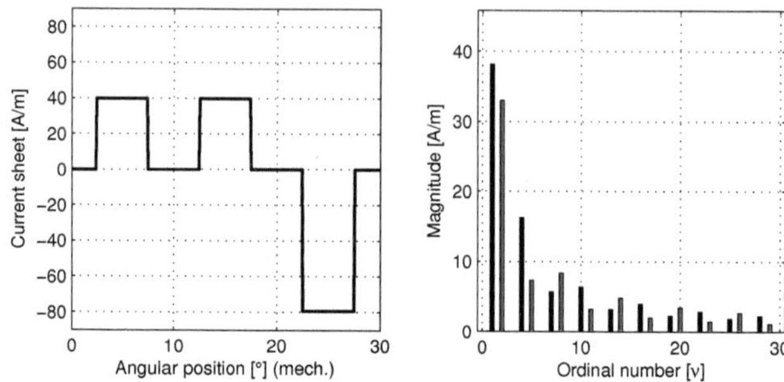

Figure 3.8.: Elementary Machine winding arrangement 2/3. black: Forward rotating wave; red: Backward rotating wave

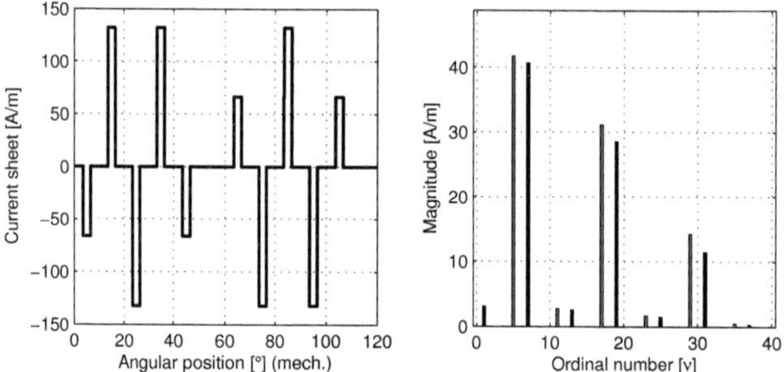

Figure 3.9.: Elementary Machine winding arrangement 10/12. black: Forward rotating wave; red: Backward rotating wave

Table 3.2.: Harmonic Orders.

Winding arrangements	Orders									
$< 3-1-1-2 >$	1	-2	4	-5	7	-8	10	-11	13	-14
$< 12-2-2-10 >$	-1/5	5/5	-7/5	11/5	-13/5	17/5	-19/5	23/5	-25/5	29/5

to higher iron losses. In figure 3.8, the harmonics black bar indicates the forward rotating wave in the direction of fundamental and the red bar indicates the wave that rotate in the opposite direction i.e against the fundamental. Since in fractional-slot winding arrangement the term q is not a whole number, the symmetry conditions should be fulfilled among the number of phases m, slots Q_s and poles $2p$. The number of slots per pole per phase is rearranged as,

$$q_k = \frac{Q_s}{2pm} = \frac{Q_s/m}{2p} = HCF \cdot \frac{z}{n} \tag{3.33}$$

The ratio z/n indicates that z coils per phase (the numerator) must be distributed among successive poles n (the denominator). The highest common factor (HCF) determines the number of repeatable sections of base elementary machine. HCF also provides information on possible parallel arrangement of the winding. There are methods to distribute the coils around the armature periphery. However, tabular construction also known as Tingley's scheme is quite simple and this approach is not discussed here. This method is explained in publications [38] and [44]. Vogt [45] distinguished fractional-slot type of winding arrangements as 1st grade and 2nd grade based on denominator value and also on layer of winding type. The harmonics order (Table 3.2) for the considered stator arrangement in figures 2.15 and 2.16 is expressed mathematically,

For 2nd grade type with even denominator in Equation,

$$\frac{\nu}{p} = \pm \frac{1}{n}(2mg + 2) \quad g = 0, \pm1, \pm2, \pm3, ... \tag{3.34}$$

For 1st grade type with odd denominator in Equation,

$$\frac{\nu}{p} = \pm \frac{1}{n}(2mg + 1) \quad g = 0, \pm1, \pm2, \pm3, ... \tag{3.35}$$

Based on these Equations 3.34 and 3.35, the order of harmonics are calculated and it is tabulated (Table 3.2). The stator armature field is given by,

$$B_s(x_s, t) = \sum_{\nu=1,3,5...}^{\infty} B_\nu cos(\nu\pi/\tau_p - \omega t) \tag{3.36}$$

where B_ν is given by,

$$B_\nu = j \cdot \frac{\mu_0 \cdot A_s}{\delta_m} \tag{3.37}$$

where δ_m is the effective air gap.

Figure 3.10 shows armature flux density for D1 geometry for both analytical and FEM calculations. The flux density is estimated at the middle of the air gap. The field distribution is the radial component for two pole periods. The field solution used analytically for this calculation is derived in Appendix B. The field distribution of winding arrangement is explained in section 3.3.3. There are hardly differences between the analytically calculation and the FEM result.

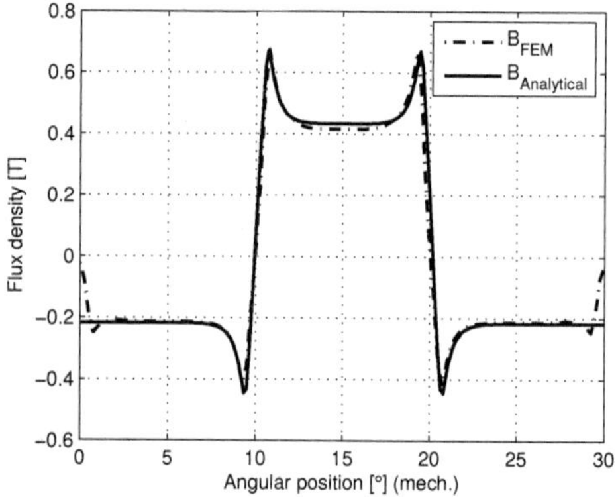

Figure 3.10.: Armature radial flux density (D1) at the middle of the air gap.

3.2.5. Permeance due to Slotting

The previously explained stator and rotor fields are derived for smooth surface in the air gap. These fields are modulated due to slot openings thus the flux density is reduced by Carter factor k_c.

$$k_c = \frac{\tau_s}{\tau_s - \zeta b_N} \tag{3.38}$$

where τ_s is the slot pitch, b_N is the slot opening and ζ is ratio of slot opening to air gap length and is defined as,

$$\zeta \approx \frac{b_N/\delta_m}{5 + b_N/\delta_m} \tag{3.39}$$

Because of the presence of slot opening the air gap permeance varies periodically around the air gap, the air gap permeance function due to slot opening is given by [46],

$$\lambda_{r,\theta} = \frac{1}{k_c}\cdot(1 - \lambda_Q cos(Q_s\theta)) \tag{3.40}$$

Now the magnetic flux density including the slot effect is expressed as,

$$B_{\delta q}(x_r, t, \gamma') = B_{\delta q}(x_r, t)\cdot\lambda_{r,\theta} \tag{3.41}$$

where γ' denotes the reference angular position with respect to the slotless field [46]. There are some methods to determine the air gap permeance function for slot opening. One method is based on conformal transformation which is mentioned for the armature field in [47] and for the PM field in [48]. Another method mentioned in [46] is done with the help of FEM tool by comparing the no-load flux density with and without slot opening. In this thesis work, the fore-mentioned method based on conformal transformation for scalar potential fields is used. Basic details of conformal transformation especially Schwartz-Christoffel transformation is mentioned in Appendix A. Using conformal transformation (CT), complicated geometries such as slot-opening can be transformed or mapped from one plane in to another plane by mathematical transformation methods and by having the field solution in that plane.

Armature Field with Slot

For the tranformation of the slot it is assumed that slot is infinitely deep. The field solution for the armature slot current is carried out in three stages of mapping as described in [47]. The steps involved for the plane transformation are $S{\rightarrow}Z{\rightarrow}W{\rightarrow}T$. The transformation is logarithimic transformation between S plane and Z plane. It is given by,

$$z = ln(s) \tag{3.42}$$

where $s = u + iv = re^{j\theta}$ and $z = x + jy$. In one slot having current of $2I$ [A] the potentials are I and $-I$ at the teeth sides and the zero potential is at the axis AA' (Figure 3.11). Due

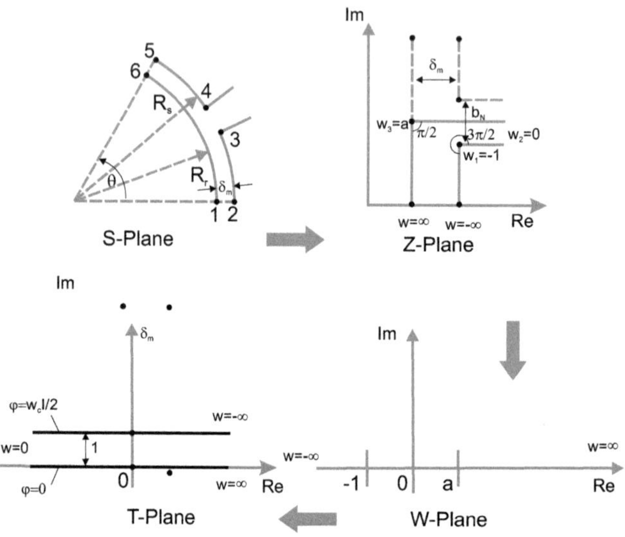

Figure 3.11.: Transformation involved in armature field.

to the symmetry along the axis AA', only one half of the slot is considered (Figure 3.11)) in the Z-plane. The interior angles of the polygon in the Z-plane at vertices w_1, w_2, w_3 are $3\pi/2, 0$ and $\pi/2$. The corresponding points in the Z-plane and W-plane are,

$$at \quad w_1 \quad z = b + j\delta_m \quad w = -1$$

$$at \quad w_2 \quad z = j \quad\quad\quad w = 0$$

$$at \quad w_3 \quad z = 0 \quad\quad\quad w = a$$

The Schwartz-Christoffel equation between Z-plane and W-plane transformation is given by,

$$\frac{dw}{dt} = A(w+1)^{\frac{3\pi/2}{\pi}-1}(w-0)^{\frac{0}{\pi}-1}(w-a)^{\frac{\pi/2}{\pi}-1} = A\frac{\sqrt{(w+1)}}{w\sqrt{w-a}} \tag{3.43}$$

On substituting in Equation 3.43,

$$p^2 = \frac{w-a}{w+a} \tag{3.44}$$

After integration,

$$Z = 2A\left(\frac{1}{\sqrt{a}}\arctan\frac{p}{\sqrt{a}} + \frac{1}{2}ln\left(\frac{1+p}{1-p}\right)\right) + C_1 \tag{3.45}$$

The constant A is determined using method of residues [47] and it is given by,

$$A = -j\frac{g'}{\pi} \tag{3.46}$$

Using this value of A and substituting for the points $w = 0$, the unknown a is expressed as,

$$a = \left(\frac{2g'}{b'_N}\right)^2 \tag{3.47}$$

The field for the slots is finally determined in the T-plane. Here the upper half of the W-plane is represented as two semi-infinite parallel planes: one between 0 and ∞, and the other between 0 and $-\infty$ separated by unit distance. These are shown in Figure 3.11. The required transformation is logarithmic function and it is given by,

$$l = \frac{1}{b\pi}lnw \tag{3.48}$$

The potential of the lower plate is at $\varphi = 0$ and the upper plate is $\varphi = -\frac{w_c}{2}I$. The field solution is then,

$$B_t = j\mu_0\frac{z_c}{2}I \tag{3.49}$$

where z_c is the number of conductors in slot. Now the field solution in the S plane is given by,

$$B_s = B_t\left(\frac{\partial t}{\partial w}\frac{\partial w}{\partial z}\frac{\partial z}{\partial s}\right)^* = -j\frac{1}{g'}\frac{\sqrt{w-a}}{\sqrt{w+1}}\frac{1}{s}B_t \tag{3.50}$$

On substitution for B_t,

$$B_s = \mu_0\frac{w_c}{2g'}I\frac{\sqrt{w-a}}{\sqrt{w+1}}\frac{1}{s} - \lambda^*B_t \tag{3.51}$$

where λ is the permeance due to slot for the armature field and it is expressed as,

$$\lambda = \frac{\sqrt{w-a}}{\sqrt{w+1}}\frac{1}{s} \tag{3.52}$$

PM Field with Slot

The steps involved for the PM field transformation are performed in four stages [48] $S{\rightarrow}Z{\rightarrow}W{\rightarrow}T{\rightarrow}K$ as shown in the Figure 3.12 . The first transformation $S{\rightarrow}Z$ is logarithmic mapping of cylindrical coordinates into cartesian coordinates. Then Z plane is transformed in to upper half of the W plane using Schwartz-Christoffel tranformation method. The $W{\rightarrow}T$ mapping is also performed with Schwartz-Christoffel tranformation method. Final tranformation $T{\rightarrow}K$ is exponential mapping. In the last tranformation the slot in the S plane is represented as slot less circular K plane in which the field is obtained. The field solution in the S plane is described as [48],

$$B_s = B_k\left(\frac{\partial k}{\partial t}\frac{\partial t}{\partial w}\frac{\partial w}{\partial z}\frac{\partial z}{\partial s}\right)^* = \lambda^*B_k \tag{3.53}$$

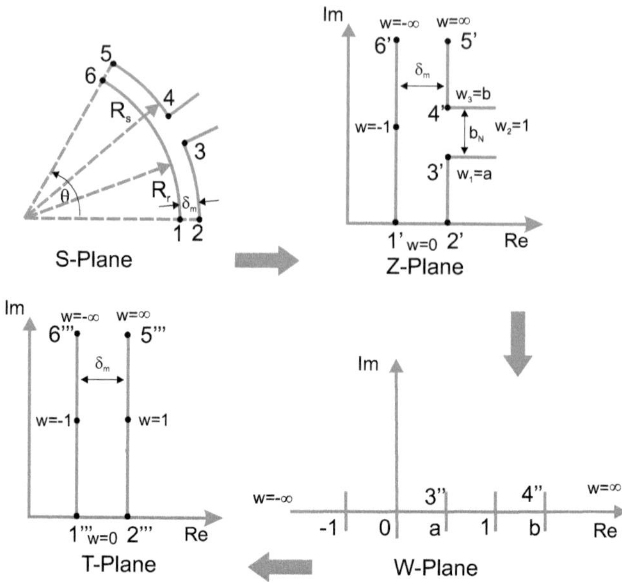

Figure 3.12.: Transformation involved in PM field.

where λ is the permeance due to slot for the PM field and it is expressed as,

$$\lambda = \frac{k}{s}\frac{(w-1)}{\sqrt{w-a}\sqrt{w-b}} \tag{3.54}$$

In this work another air gap permeance function described in [41] is also used to determine the forces at the lateral sides of the slot opening. It is calculated with the assumption shown in the Figure 3.13. The magnetic field distribution at the air gap is assumed that flux crosses in straight path from the magnet to the tooth and its path is circular at the slot opening having tooth tip corner as center. The function for this permeance over one slot pitch is represented as,

$$f(x) = \lambda' \quad \text{for } \tau_s - b_N/2 < x < \tau_s + b_N/2 \ .$$

Thus the relative air gap permeance at the slot opening mathematically expressed as,

$$\lambda' = \frac{\lambda}{\left(\frac{\mu_0}{g+h_m/\mu_r}\right)} \tag{3.55}$$

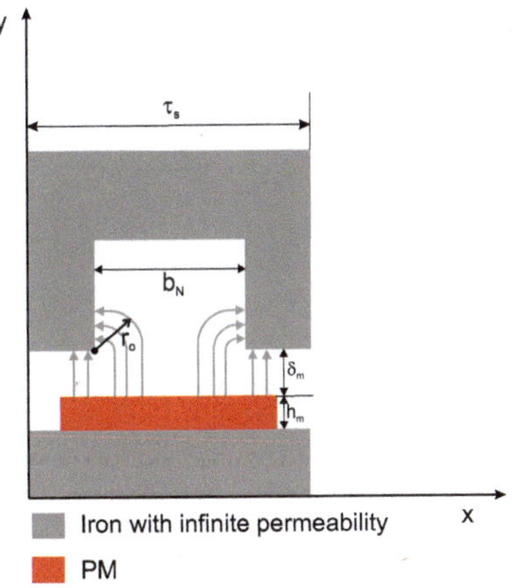

Figure 3.13.: Air gap permeance PM field.

where λ is given by,

$$\lambda = \frac{\mu_0}{g + \frac{h_m}{\mu_r} + \frac{2\pi r_o}{4}} \tag{3.56}$$

The slot permeance is represented as general form of Fourier series,

$$\lambda_s(r,\theta) = \lambda_{s,avg}(r) + \sum_n \lambda_n cos(nQ_s\theta) \tag{3.57}$$

where Q_s is the number of slots. The no-load flux density with air gap permeance calculation is discussed in the section 3.2.8.

3.2.6. Permeance due to Saturation

Due to the non-linearity in the material the electrical machines are subjected to saturation under load and also under no-load. This can be conveniently characterized as permeance function ([50] and [51]). The saturation occurs at one elementary machine usually due to the fundamental component i.e. with every $2p$ positions and travels at angular velocity ω/p same as fundamental. Mathematically it is expressed as,

$$\lambda_{sat}(r,\theta) = \lambda_{sat,avg}(r) + \sum_n \lambda_n cos2n(p\theta - \omega t) \tag{3.58}$$

where $n > 1$ are applicable for the heavily saturated machine. The mean air gap permeance due to saturation is written as [52] [53],

$$\lambda_{sat,avg} = \frac{\pi}{2k_\alpha} \cdot \frac{\mu_0}{\delta_m} \cdot \frac{1}{k_{sat}} \tag{3.59}$$

where k_α is the factor due to flattening of air gap flux density. The saturation factor k_{sat} can also be determined using magnetic voltages,

$$k_{sat} = \frac{\sum U_{s,r}}{U_\delta} \tag{3.60}$$

where $U_{s,r}$ is the magnetic voltages in the stator and rotor and U_δ is the magnetic voltage in the air gap. Alternatively, the saturation factor is calculated with the help of FEM program as described in [46]. Two linearly calculated no-load flux density and armature flux density $(B_s + B_r)$ are superimposed. The reduction of peak field between directly calculated field at the air gap due to load current and the sum of the stator and rotor fields $(B_s + B_r)$ gives the saturation factor. All the ordinal numbers in the stator armature field are multiplied with reciprocal of saturation factor to have the influence of saturation.

The flux density due to saturation is given by,

$$B_{sat}(x_r, t) = B_s(x_s, t) \cdot \lambda_{sat}(r, \theta) \tag{3.61}$$

Because of the saturation current loading will magnetize some parts of the magnetic circuit than magnetizing the air gap alone. Due to saturation, influence air gap flux density curve is flattened and thus harmonics of order $3p$ is produced.

3.2.7. Permeance due to Magnets

If the magnet relative permeability is considered same as air, the air gap permeance is given by,

$$\Lambda_{air} = \frac{\mu_0}{g + h_m} \tag{3.62}$$

When relative permeability of the magnet is considered then the effective air gap is calculated by assuming simple magnetic circuit [43]. Then the air gap permeance is given by,

$$\Lambda_{air} = \frac{\mu_0}{g + \frac{h_m}{\mu_r}} \tag{3.63}$$

The air gap permeance variation in the circumference is then according to [49] and [54],

$$\lambda_{pm} = \lambda_{pm,avg} + \sum_{n=1}^{\infty} \lambda_{pm,n} cos(2np(\theta - \omega t)) \qquad (3.64)$$

The constants $\lambda_{pm,avg}$ can also be calculated with the help of FEM [46].
The resultant air gap flux density is represented as,

$$b(\theta, t) = B_{sum}(\theta, t) \cdot \lambda_{sum}(\theta, t) \qquad (3.65)$$

where $\lambda_{sum}(\theta, t)$ is,

$$\lambda_{sum}(\theta, t) = \lambda_s(r, \theta) + \lambda_{sat}(r, \theta) + \lambda_{pm}(\theta, t) \qquad (3.66)$$

where $\lambda_s(\theta, t)$ includes both slot permeance of the armature field and the PM field. The permeance variation due to eccentricity is not included in this work because eccentricity permeance variation is interesting for radial forces and noise calculations ([55]). If the machine has eccentricity and the stator winding arrangement contains parallel paths then induced emf generated is unequal and generates circulating currents. Steinbrink [21] made experiments of machine with eccentricity and results for torque pulsation is discussed in the work. The $B_{sum}(\theta, t)$ includes the sum of armature and PM field and it is represented as

$$B_{sum}(\theta, t) = B_{stator}(\theta, t) + B_{rotor}(\theta, t) \qquad (3.67)$$

The mathematical expression for the total fields in analytical form contains all the above mentioned components. The interaction of these combinations are responsible for parasitic tangential forces which is going to be explained in the later section.

3.2.8. No-load Flux Density with Slot

The geometry D1 is considered for the analytical calculation of no-load flux density. The slot opening width is taken as 5 mm and slot opening depth is 2.5 mm. The magnet pole coverage is considered as 0.67. Other data used for the calculation are provided in the Table 3.1. The number of ordinal numbers considered for Fourier series of flux density and permeance function is 50. The plot of no-load flux density (radial and tangential) at the middle of the air gap is shown in Figure 3.14. As expected, there are two dips for the $< 3 - 1 - 1 - 2 >$ winding arrangement for two pole periods. The calculations are made for two different analytical results,

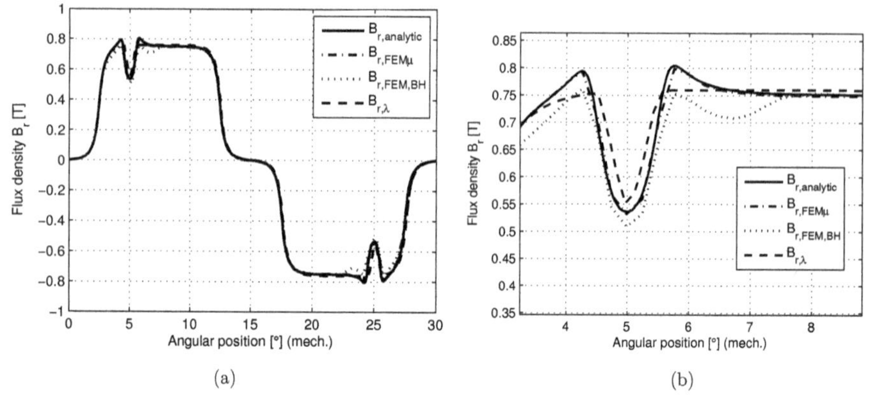

(a)   (b)

Figure 3.14.: No-load radial flux density (D1). (a)Two pole period. (b)Zoomed at dip.

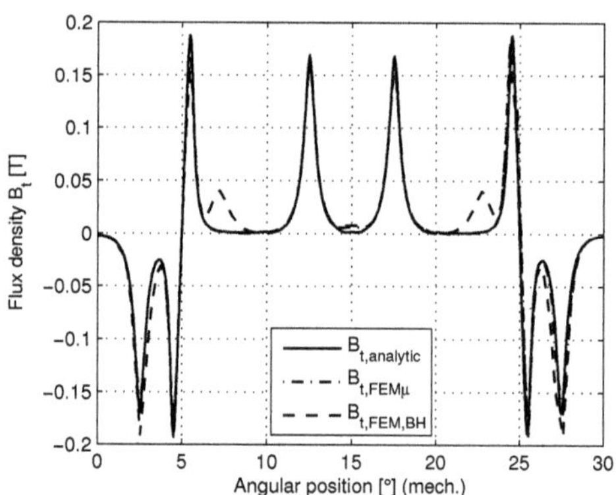

Figure 3.15.: No-load tangential flux density (D1).

one with the help of permeance function based on conformal mapping $B_{r,analytic}$ using [48] and the other permeance function based on Equation 3.56 $B_{r,\lambda}$. FEM calculations have been done for linear material with constant relative permeability $B_{r,FEM\mu}$ and also for non-linear material $B_{r,FEM,BH}$. The comparison of the no-load flux density gives the quality of the analytical calculation (refer Figure 3.14). Due to the saturation in the tooth tips, the non-linear material calculation from FEM calculation has additional dip near the slot-opening. The flux density distribution in other positions from analytical results have excellent agreement with the FEM calculations. The same comparison has been made for tangential component and it is shown in Figure 3.15.

3.3. Torque Derivation

Torque or force derivation for electrical machines is based on Maxwell stress tensor, Energy methods and Lorentz force. Maxwell stress tensor and Lorentz force methods related to this work are discussed. The volume force is defined as,

$$\iiint_V f \cdot dv = \iiint_V (f_e + f_b) \cdot dv \tag{3.68}$$

This equation contains force due to both electric f_e and magnetic f_b terms. Consequently this force density represents the Lorentz force and it takes the form,

$$\vec{F} = \iiint_V q(\vec{E} + v \times \vec{B}) \tag{3.69}$$

The first term is the electric field strength and it does not contribute to force since there is no free charge. From Lorentz equation $\vec{J} \times \vec{B}$ and using Ampere's law $\nabla \times \vec{H} = \vec{J}$ and taking magnetic flux density $\vec{B} = \mu_0 \mu_r \vec{H}$ in air gap the Equation 3.69 is written as the force density f_v ,

$$f_v = \left(\nabla \times \frac{\vec{B}}{\mu_0} \right) \times \vec{B} \tag{3.70}$$

This vector products component can be derived in to tensors and the volume force density is given by [7],

$$\tau_v = \nabla \cdot T \tag{3.71}$$

where T is expressed as,

$$T = \frac{1}{\mu_0}\begin{bmatrix} B_x^2 - \frac{1}{2}|B|^2 & B_xB_y & B_xB_z \\ B_yB_x & B_y^2 - \frac{1}{2}|B|^2 & B_yB_z \\ B_zB_x & B_zB_y & B_z^2 - \frac{1}{2}|B|^2 \end{bmatrix} \tag{3.72}$$

The volume force density is given by,

$$\vec{F} = \iiint_V \nabla\cdot T \tag{3.73}$$

Using Gauss divergence theorm, the integral of divergence of volume $\vec{F}$ is equal to the surface integral of the vector $\vec{F}$. The maxwell stress tensor which is used to calculate the force is defined as the force on an enclosed body obtained by integrating radial and tangential component of the tensile stress around the contour of closed surface. It is given by,

$$\tau_w = \frac{1}{\mu_0}(\vec{B}\cdot\vec{n})\vec{B} - \frac{1}{2\mu_0}|B|^2\cdot\vec{n} \tag{3.74}$$

where $\vec{n}$ is the unit normal vector. Usually in this approach, the boundary is selected at the air gap region. If the tangential and radial components of the flux density is available, then the force can be calculated.

Reluctance Torque

In the Equation 3.74 an additional term should be considered which is responsible for the force due to change of permeability of the material at the interface. This force is termed as reluctance force and it is defined according to [52] and [21],

$$\tau_x = -\frac{1}{2\mu_0}|H|^2\nabla\mu \tag{3.75}$$

In the macroscopic level there exists tensile stress along the lines of force and compressive stress at right angle to the lines of force. The former tends to shorten the magnetic flux and the later tends to widen the magnetic flux [52]. Based on the materials at the interface two theories from Carter and Helmholtz materials are discussed in [56]. The direction of force is pointed from higher magnetic permeability to the lower magnetic permeability. In electrical machines, Helmholtz's material assumption is valid and the force referred to the rotor coordinate system has only tangential component in the slot edges at the interface between iron and air because iron is assumed as $\mu \to \infty$, thus the normal magnetic field is zero. The expression for tangential force density at the slot edges based on Helmholtz is written as,

$$\tau_x = \frac{1}{2\mu_0} B_{\delta,x}^2 \tag{3.76}$$

where $B_{\delta,x}$ is the tangential flux density at the slot sides. This tangential flux density is considered as $B_{\delta,r}$ the radial flux density at the air gap because the magnetic flux which is $\Phi = \int B \cdot ds$ according to Gauss law is conserved, i.e surface integral along the closed surface is zero. The considered assumption for the reluctance force calculation is shown in the Figure

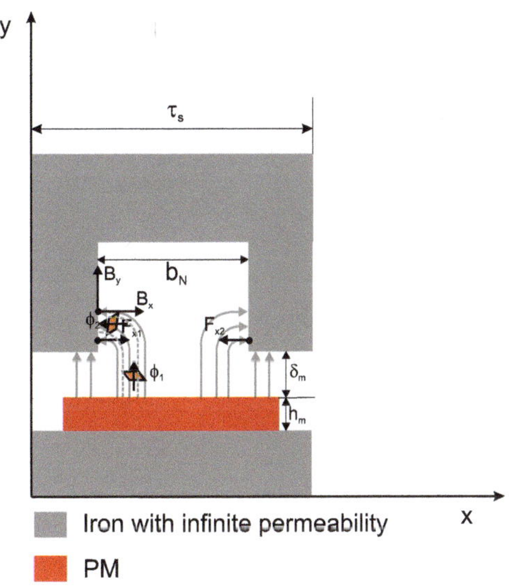

Figure 3.16.: Slot edge field and force.

3.16. The average torque can be calculated at the air gap radius r for the machine length l. It is given by,

$$M = F \cdot r \tag{3.77}$$

where total force F is,

$$F = 2\pi \cdot r \cdot l \cdot (\tau_w + \tau_x) \tag{3.78}$$

3.3.1. Mathematical Formulation of Cogging Torque

Cogging torque occurs under no-load due to the interaction of magnetic field harmonics and air gap permeance harmonics. Based on Maxwell stress tensor the torque can be written as,

$$M = \frac{1}{\mu_0} r^2 \cdot l \int_0^{2\pi} B_r B_\theta \cdot d\theta \qquad (3.79)$$

where the flux density components B_r and B_θ are in Equations 3.27 and 3.29. These flux density components are modulated with air gap permeance function Equation 3.66. The resultant flux density components are periodic functions of sine and cosine.

The tangential forces are responsible for the torque production. The rules for generation of torque [57] are understood from the Equation 3.80: The term yield zero when $\mu \neq \nu$,

$$M = \frac{1}{\mu_0} r^2 \cdot l \cdot B_{r,\mu} B_{\theta,\nu} \int_0^{2\pi} \sin(\mu\theta)\cos(\nu\theta) \cdot d\theta = 0 \qquad (3.80)$$

This term is non-zero only for certain argument combinations. Therefore torque exists for certain orders of harmonics of the radial flux density component $\mu = n$ when same order of harmonics is available in the tangential flux density component $\nu = n$. For all other combination torque is zero. It means that torque will be generated by interaction of two sinusoidal fields (rotor and stator) in the air gap only by harmonics with identical pole-pairs of two interacting fields. Before considering an example some trigonometry expressions are provided in the Equations 3.81 and 3.82,

$$2\sin A\cos B = \sin(A+B) + \sin(A-B) \qquad (3.81)$$

$$2\cos A\cos B = \cos(A+B) + \cos(A-B) \qquad (3.82)$$

The fundamental space harmonics of radial flux field with slot harmonics,

$$B_r = \hat{B}_\mu cos(p\theta - \omega t) \cdot \lambda_s cos(nQ_s\theta)) \qquad (3.83)$$

where Q_s is the number of slots. Using trigonometric expansion for the above equation,

$$B_r = \frac{\hat{B}_\mu \lambda_s}{2}[cos((p+nQ_s)\theta - \omega t) + cos((p-nQ_s)\theta - \omega t)] \qquad (3.84)$$

Similar expression for the tangential component is,

$$B_t = \frac{\hat{B}_\nu \lambda_s}{2}[-\sin((p+nQ_s)\theta - \omega t) - \sin((p-nQ_s)\theta - \omega t)] \tag{3.85}$$

Consider an elementary machine $< 3-1-1-2 >$ for $Q_s = 3$ slots and $p = 1$, the first argument in Equations 3.84 and 3.85 have ordinal number (-2,4,-5,7,...). On making substitutions of Equations 3.84 and 3.85 in Equation for Torque 3.80.

For the 5th harmonics combination the Equations 3.84 and 3.85 are expressed as,

$$M(\theta) - -\frac{1}{\mu_0}r^2 \cdot l\ B_r B_\theta \int_0^{2\pi} \sin(5\theta + \omega t)\cos(5\theta - 5\omega t)\cdot d\theta \tag{3.86}$$

Let $X1 = (5\theta + \omega t)$ and $X2 = (5\theta - 5\omega t)$. Using equations 3.81 and 3.82, $\sin(X1 + X2)$ term goes to zero and $\sin(X1 - X2)$ is the non-zero component. After solving the Equation 3.86, the torque is

$$M(\theta) = -\frac{1}{\mu_0}r^2 \cdot l \cdot B_r B_\theta \sin(6\omega t) \tag{3.87}$$

For the 7th harmonics combination the Equation 3.80 is expressed as,

$$M(\theta) = \frac{1}{\mu_0}r^2 \cdot l \cdot B_r B_\theta \int_0^{2\pi} \sin(7\theta + \omega t)\cos(7\theta - 7\omega t)\cdot d\theta \tag{3.88}$$

Now $X1 = (7\theta + \omega t)$ and $X2 = (7\theta - 5\omega t)$. With the help of equations 3.81 and 3.82, $\sin(X1 + X2)$ term goes to zero and $\sin(X1 - X2)$ is the non-zero component. After solving the Equation 3.88, the torque is

$$M(\theta) = \frac{1}{\mu_0}r^2 \cdot l \cdot B_r B_\theta \sin(6\omega t) \tag{3.89}$$

For this example the cogging torque pulsates at six times the fundamental frequency which corresponds to six pulsations in one elementary machine for $< 3-1-1-2 >$ arrangement. As shown in Chapter 4, there are different combinations among the harmonics. The total torque is sum of all the combinations. The results of cogging torque is discussed in next section 3.3.2. As shown in Figure 1.2, the prototype paper mill machine is not so easy to rotate with free hand. This is due to increased slot permeance and the rotor field tends to align with stator slot openings. This phenomenon is a typical cogging torque for permanent magnet machine. The Figure 3.17 points direction of the forces $\vec{F}_{xi}; i = 1, 2, 3, 4$ i.e. from higher permeability to the lower permeability for 2 slots and a magnet arrangement [56]. At position A (3.17a) all

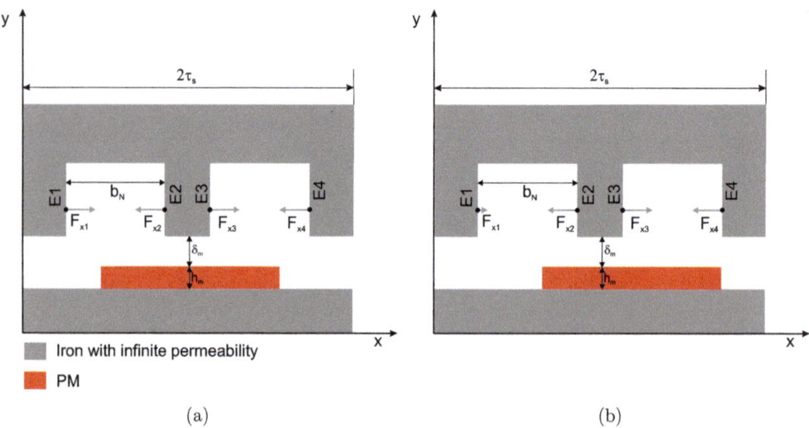

Figure 3.17.: Tangential Force. (a) Position A. (b) Position B.

the forces at slot sides are equal. The net force is zero and so is the cogging torque. If the magnet is moved to position B (3.17b), then the field on the slot sides changes and the forces also change. The force vectors $\vec{F}_{x,3}$ and $\vec{F}_{x,4}$ are increased and the force vectors in the other slot edges are decreased. The slot edges $1, 3$ tend to pull the magnets until the field entering is maximum and the direction of force changes when the field is minimum. The net cogging torque for one slot is then,

$$M = \frac{1}{2\mu_0} r{\cdot}l \int_0^{b_N/2} (B_{x1}^2 - B_{x2}^2){\cdot}dR \tag{3.90}$$

where B_{x1} and B_{x2} are the tangential fields in the slot edges $E1 \& E2$ and b_N is the slot opening. B_{x1} generates force F_{x1} on the left side of the slot and B_{x2} generates force F_{x2} on the right side of the slot. As explained earlier, this method which is also used in this work determines the cogging torque at slot sides based on Helmholtz material assumption. The calculation based on this method is simple to apply because only the magnitude of the air gap flux density is required. Moreover iron material is assumed with infinite permeability, and only the radial component is enough to determine the torque.

3.3.2. Cogging Torque Result Comparison

The analytically calculated cogging torque results are validated with FEM calculation. The calculation is carried out for the machine model D1 and for the parameters provided in Table 3.1. The slot opening width is $5mm$ and slot opening depth is $2.5mm$. The results are compared for two analytical methods with FEM. In FEM, a linear material with high permeability for iron ($M_{FEM,\mu}$) and a non-linear material ($M_{FEM,BH}$) are considered. The curve which is shown in Figure 3.18 is for one slot pitch and for mechanical angle of $5°$ for D1 geometry. In Figure

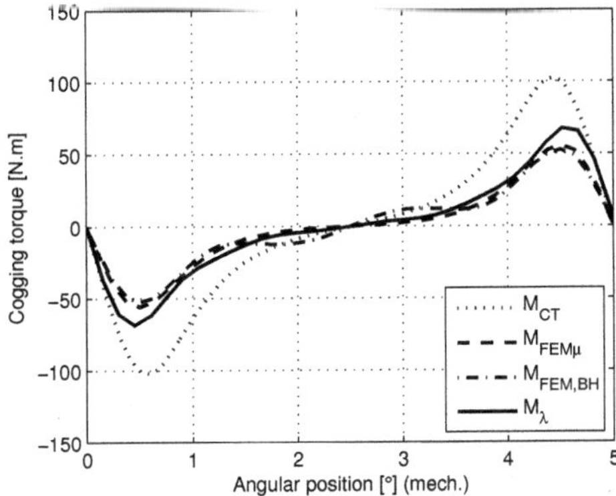

Figure 3.18.: Cogging Torque comparison (D1).

3.18, only the two curves obtained from FEM calculation are compared, there is small variation because of even under no-load. This is due to the tooth tips which are saturated. This changes the air gap permeance and also influences the cogging torque. The analytically calculated torque curve M_λ based on slot sides (Helmholtz) method is very close to FEM results. Only the amplitude varies because the value of permeance are determined from simple slot geometry parameters. There is a considerable amount of difference in amplitude from the method based on maxwell stress tensor using conformal mapping (M_{CT} using [49]). The result from the (M_{CT}) which uses conformal transformation is as expected and it is explained using Figure 3.19.

In the Schwartz-Christoffel plane transformation in the conformal mapping the teeth edge is transformed into a plane which is a straight line. The cogging torque value is calculated mathematically after the transformation. Whereas in FEM, the mesh which is at the tooth tip, has its average potential values away from the tooth tip since there is no point mesh. However,

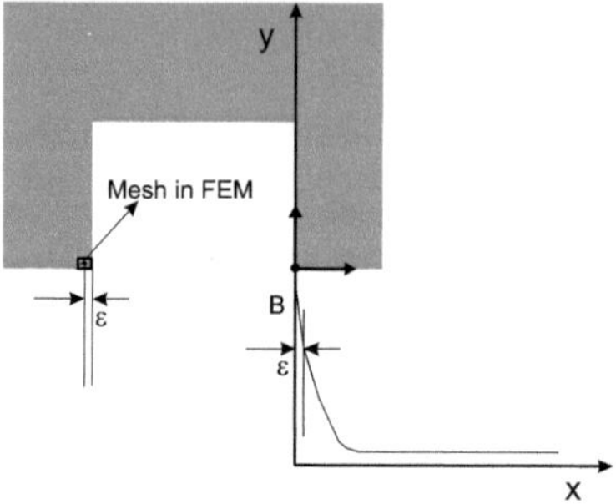

Figure 3.19.: Cogging Torque comparison (D1).

Table 3.3.: Cogging torque comparison (D1).

Definition	Peak-Peak Torque	Units
Conformal Mapping	101.8	$N.m$
Slot sides (Helmholtz)	67.4	$N.m$
FEM linear material	55.7	$N.m$
FEM non-linear material	52.5	$N.m$

the torque which is calculated analytically can also be adjusted to the FEM values if the field point is moved away from the tooth edge with the variable ε as shown in Figure 3.19. Starschich [56] and Zarko [48] used auxillary parameter to overcome the singularity at the tooth tip during conformal transformation for Helmholtz method.

The peak values of the cogging torque from different approaches is given in Table 3.3. In all the methods the periodicity is kept similar for D1 geometry. The torque based on slot sides (Helmholtz) method is nearer to the FEM calculated values. The FEM calculation will be most close to reality because the non-linear material properties is considered for the calculation. The analytical methods are able to predict the cogging torque very faster than FEM calculation and it assists the FEM calculation at the initial design stage.

3.3.3. Mathematical Formulation of Torque Ripple

The force exerted on a conductor is calculated using Lorentz force method as

$$\vec{F} = \iiint_V (\vec{J} \times \vec{B}) \tag{3.91}$$

where $\vec{J}$ is the current density and $\vec{B}$ is the flux density. The $\vec{J}$ and $\vec{B}$ are composed of harmonics and also modulated with permeance functions. Again the rule mentioned in Equation 3.80 says that torque pulsation exists for only certain harmonics combination. Now consider an example, the stator current sheet is given by,

$$A_s = \sum_{\nu}^{\infty} \hat{A}_\nu \sin(\nu p\theta - \omega t) \tag{3.92}$$

where ν is given in the Table and A_ν is calculated according to the Equation 3.32. The rotor field without slot is expressed as,

$$B_r = \sum_{\mu}^{\infty} \hat{B}_\mu \cos(\mu p\theta - \mu\omega t) \tag{3.93}$$

where μ is given in Equation 3.30. For 5th harmonics combination from the stator current sheet and the rotor field, the equation is expressed as

$$M(\theta) = \frac{1}{\mu_0} r^2 \cdot l \cdot A_s B_r \int_0^{2\pi} \sin(5\theta + \omega t)\cos(5\theta - 5\omega t) \cdot d\theta \tag{3.94}$$

Let $X1 = (5\theta + \omega t)$ and $X2 = (5\theta - 5\omega t)$. Using Equation 3.82, $sin(X1 + X2)$ term goes to zero and $sin(X1 - X2)$ is the non-zero component. After solving the Equation 3.94, the torque is

$$M(\theta) = \frac{1}{\mu_0} r^2 \cdot l \cdot A_s B_r \sin(6\omega t) \tag{3.95}$$

In an elementary machine for $< 3 - 1 - 1 - 2 >$ arrangement if the dominating components of 5th and 7th ordinal number exists then the torque pulsation will have six periods. The total torque is sum of cogging torque and electromagnetic torque (Equation 3.78). The minimum torque ripple pole coverage is not the same as minimum cogging torque pole coverage [58]. Usually, the pole coverage of the magnets can be selected according to the winding harmonics [59]. The half-range Fourier series of permanent magnet is given by,

$$B_\delta(x_r, t) = \sum_{n=1,3,5...}^{\infty} \frac{B_r}{\mu_0} \frac{4}{n\pi} sin\left(\frac{n\pi\alpha_p}{2}\right) \cdot cos\left(\frac{\nu\pi}{\tau_p}\right) \tag{3.96}$$

where α_p is the magnet pole coverage. The winding factor of $< 3 - 1 - 1 - 2 >$ and $< 12 - 2 - 2 - 10 >$ machine arrangements are shown in Figure 3.20. For one elementary machine

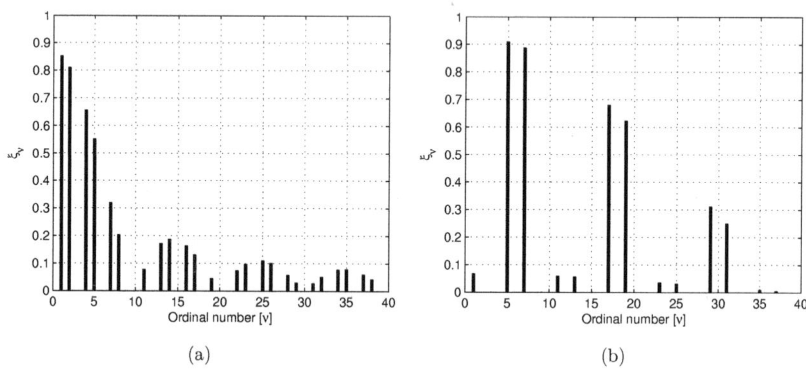

Figure 3.20.: Winding factor. (a) $< 3 - 1 - 1 - 2 >$ (b) $< 12 - 2 - 2 - 10 >$

of $< 3 - 1 - 1 - 2 >$ arrangement the winding factor of 5th harmonics is considerably large (0.55). The Equation 3.96 value is zero when α_p $(2/5, 4/5, 8/5, \ldots)$ is selected. Thus torque pulsation due to 5th harmonics can be reduced with pole coverage of $4/5$. If α_p selected as $(2/7, 4/7, 8/7, \ldots)$ then torque pulsation due to 7th harmonics can be reduced.

The distribution of coils according to Tingley scheme ([38],[44]) for $< 12 - 2 - 2 - 10 >$ double layer arrangement is discussed now. The term t is obtained as,

$$t = HCF(Q_s, 2p) \tag{3.97}$$

where HCF is the Highest Common Factor. The slot angular pitch is

$$\gamma' = 180° \frac{2p}{Q_s} = 150° \tag{3.98}$$

γ' is in electrical degrees. In the Table 3.21 the number of rows is the number of poles (10 in this example) in the elementary machine and the number of columns is $\frac{Q_s}{t}$. The slot pitch is then $\frac{2p}{t}$ columns. As shown in Figure 3.21(a) for 10 poles, one starts from left/top proceeding towards right side filling up the table for $\frac{2p}{t}$ slot pitch. Figure 3.21(b) is reduced to two poles. As seen in Figure 3.21(c) coils are distributed among successive poles for one layer. The other layer is derived by coil pitch and changing the sign of the coil. In Figure 3.21(d) the winding arrangement for double layer is depicted. This winding information is used in the program for

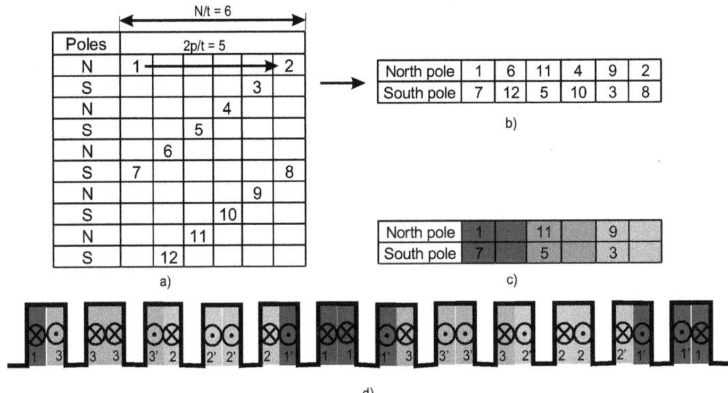

Figure 3.21.: (a)Tingley's method for $< 12 - 2 - 2 - 10 >$ arrangement, (b) Reduced poles (c, d) Reduced coil groups for two zone and double layer arrangement [38].

the analytical calculation. The maximum torque is obtained when stator field is $\pi/2$ electrically phase shifted from rotor field, i.e. stator armature current sheet and rotor field are in-phase for surface mounted synchronous machine without the reluctance torque.

3.3.4. Torque Ripple Result Comparison

Similar to cogging torque calculation as discussed in section 3.3.2, analytically calculated torque ripple result is validated with FEM calculation. The slot opening width is $5mm$ and slot opening depth is $2.5mm$ for magnet pole coverage 0.67. In FEM, torque ripple is calculated for two different lamination materials: a linear isotropic material with high constant permeability of $\mu_r = 5000$ and a non-linear material of type $M400 - 50A$. Torque ripple is calculated for load current of $176.8A$ effective. Figure 3.22 shows validation of the analytical calculation for torque ripple with FEM results for rated current. In Figure M_{FEM} represents the torque based on FEM calculation using non-linear material, $M_{FEM\mu}$ represents the torque based on FEM calculation using linear material and $M_{analytic}$ represents analytically calculated torque.

In all the cases, for one elementary machine the periodicity of the torque pulsation is six. Thus for the whole machine the torque pulsation is $p\cdot 6$. In Figure 3.22, the curves from analytical calculation and linear magnetic permeability show very good coincidence. Due to the non-linearity of the material, the torque of the machine is reduced and there is also a phase shift in the curve.

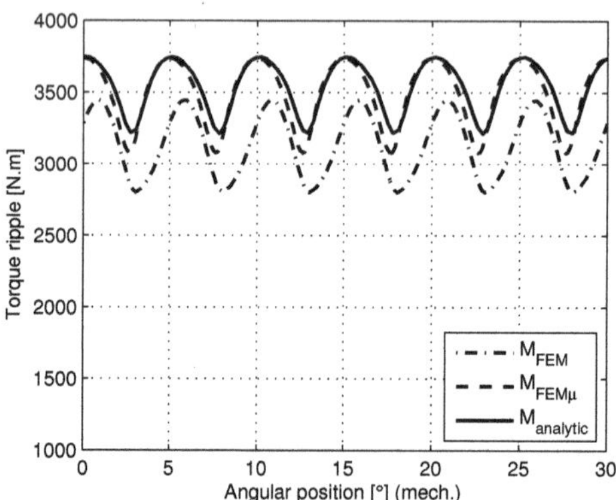

Figure 3.22.: Torque ripple comparison $@I = 176.8A$(D1).

To verify the influence of saturation on analytical calculation, the machine is calculated for half the rated current. Figure 3.23 shows the comparison of results for load current of $88.4A$. Figure 3.23 clearly depicts that for half the rated current all the curves are nearly the same.

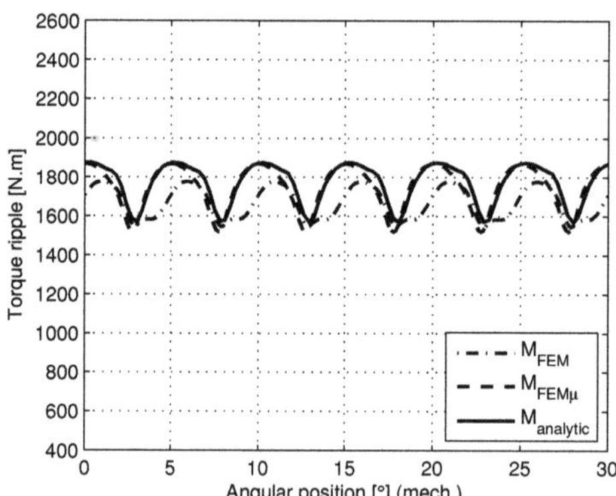

Figure 3.23.: Torque ripple comparison $@I = 88.4A$(D1).

There is small difference in the peak value of torque when comparing analytical with non-linear permeability on FEM. It should be noted that the analytical calculation yields the highest values, so it is on the safe side for the most design aspects.

Table 3.4.: Torque ripple comparison @$I = 176.8A$ (D1).

Definition	Peak-Peak Torque	Average	Units	Ripple %
Analytic	3746.9/3214.8	3544.5	$N.m$	15
FEM $\mu_r = 5000$	3737.3/3079.6	3492.4	$N.m$	18.83
FEM non-linear material	3445.3/2802.5	3128	$N.m$	20.55

Table 3.5.: Torque ripple comparison @$I = 88.4A$ (D1).

Definition	Peak-Peak Torque	Average	Units	Ripple %
Analytic	1875.2/1567.5	1774.3	$N.m$	17.34
FEM $\mu_r = 5000$	1868.6/1518.3	1746.2	$N.m$	20.1
FEM non-linear material	1777/1545.2	1665.9	$N.m$	13.9

The relative cogging torque and torque ripple in percentage is expressed as,

$$M_\% = \left(\frac{M_\Delta}{M_N}\right) \cdot 100 \tag{3.99}$$

where M_Δ is the difference between maximum and minimum value, M_N is the rated torque. The results of the torque ripple calculations are given in the Table 3.4 and 3.5.

For the rated current, the average torque is reduced but the peak-peak value is increased in FEM calculation with non-linear material assumption when compared with linear material assumption (both in analytical and FEM). For half the rated current, both average torque and torque ripple is reduced for non linear material assumption. In the next section, the influence of saturation in analytical calculation is explained.

3.3.5. Torque with Saturation Permeance

The correction of air gap permeance function because of influence of saturation in the machine is considered as in Equation 3.58. The saturation factor is determined by comparing the two superimposed linear stator and rotor fields with total field in the air gap from FEM. Figure 3.24 shows the air gap field under load and its harmonics. The fundamental of the superposition of linear field is higher. The calculated value of the saturation factor is $k_s = 0.85/0.78 = 1.1$ for the rated current of 176.8A. The analytically calculated torque ripple with saturation permeance

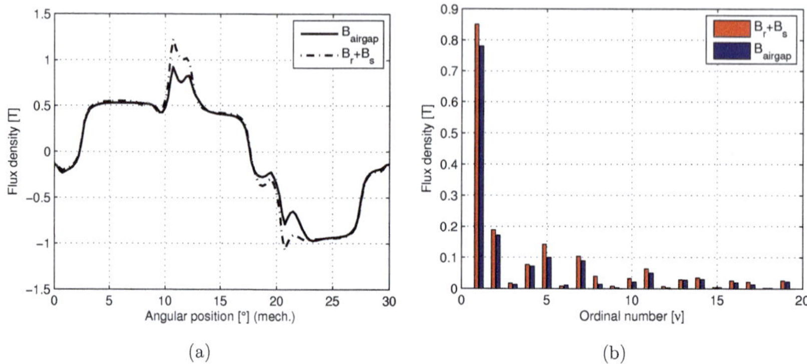

Figure 3.24.: Air gap field with load.

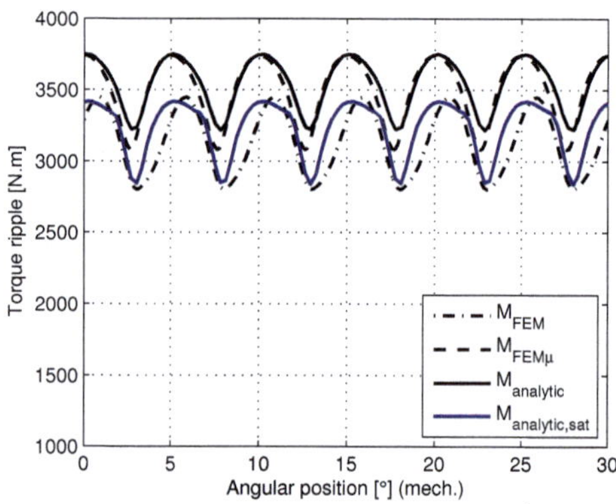

Figure 3.25.: Torque ripple comparison with saturation @$I = 176.8A$(D1).

is plotted in Figure 3.25. The local correction of saturation cannot be modelled exactly in analytical method. Thus the phase shift of the curve is not expected in the analytical model as in FEM results. The blue curve in Figure 3.25 is the analytical calculation with saturation factor. Table 3.6 shows that there are small differences in the peak to peak value between analytical calculation with saturation and FEM with non-linear material. The average value of analytical calculation with saturation is 4% more than the FEM with non-linear material. The saturation factor is calculated based on magnetostatic modelling using FEM. It means that to determine the saturation factor it is necessary to build the geometry model in FEM.

Table 3.6.: Torque ripple comparison @$I = 176.8A$ (D1).

Definition	Peak-Peak Torque	Average	Units	Ripple %
Analytic	3746.9/3214.8	3544.5	$N.m$	15
Analytic with saturation	3413.5/2834.5	3229.2	$N.m$	15
FEM $\mu_r = 5000$	3737.3/3079.6	3492.4	$N.m$	18.83
FEM non-linear material	3445.3/2802.5	3128	$N.m$	20.55

Figure 3.25 also shows that for the torque pulsation due to saturation there is a phase shift and small difference in the peak value of the curves. **But it does not produce any new harmonic order in the torque ripple waveform for surface mounted permanent magnet machine.**

3.4. Torque pulsation of Geometries

Cogging torque and torque ripple of all the geometries are discussed here. The calculations are performed in FEM. As discussed in the chapter 3.3.2, the reason is due to the consideration of non-linear material properties.

3.4.1. Cogging Torque

The waveform of the cogging torque for two pole pairs of considered geometries from D1 to D6 is shown in the Figure 3.26. The pole coverage for all the geometries is approximately 0.67, slot opening width is $5mm$ and slot opening depth is $2.5mm$. But for shell type rotor and pole-cap rotor the magnet volume is increased to maintain the same flux density as surface mounted machine. This effect is already discussed in the section 3.2.3. Geometries D1 to D3 have six periods and geometries D4-D6 have 12 periods for two pole-pair. The number of periods for an elementary machine should be multiplied by number of pole pairs p to get the frequency or periods of cogging torque for the complete machine. The peak to peak value of torque is given in Table 3.7. In Figure 3.26(a) geometry D3 has minimum peak to peak value and reduced cogging torque. In Figure3.26(b) geometry D6 has almost negligible cogging torque. When comparing surface mounted types D1 and D4, D4 has reduced cogging torque. The detailed

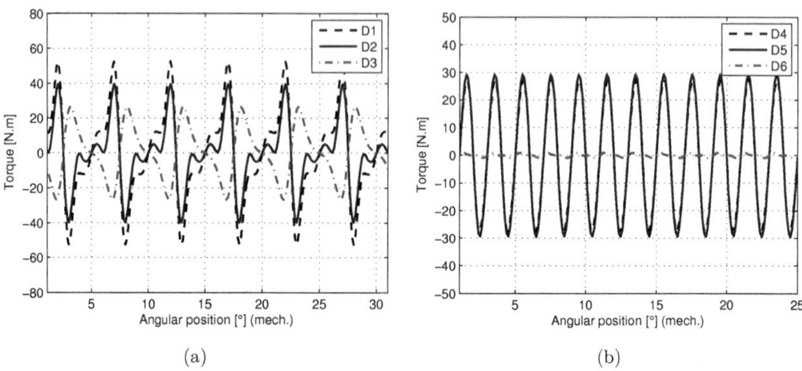

Figure 3.26.: Cogging torque for two pole-period. (a) D1/D2/D3. (b) D4/D5/D6.

Table 3.7.: Cogging torque comparison Geometries (D1 to D6).

Definition	Peak-Peak Torque	Units	Ripple %
D1	52.7/49.8	$N.m$	3.34
D2	39.6/39.6	$N.m$	2.33
D3	23.4/23.5	$N.m$	1.58
D4	27.3/27.3	$N.m$	1.72
D5	29.3/29.3	$N.m$	1.7
D6	0.9/0.9	$N.m$	0.05

interpretation based on harmonics for the generation of cogging torque is discussed in Chapter 4. As mentioned in chapter 1, the slot-pole combinations with large LCM will have reduced cogging torque.

3.4.2. Torque Ripple

Figure 3.27 depicts the waveform of torque ripple for two pole period for all the geometries. The same geometry parameters (slot opening, slot depth and pole coverage) which is taken for the coggingtorque, are considered for the torque ripple calculation. The stator armature current is $I = 176.8A$ used for the calculation. The average torque of D1 and D4 are less because the fundamental no-load flux density is less. Although the magnets of D2,D3,D5 and

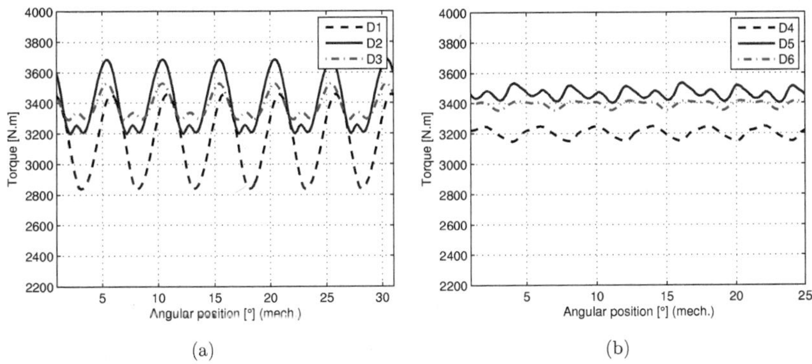

Figure 3.27.: Torque ripple for two pole-period. (a) D1/D2/D3. (b) D4/D5/D6.

Table 3.8.: Torque ripple @$I = 176.8A$ (D1).

Definition	Peak-Peak Torque	Average	Units	Ripple %
D1	3455.7/2839.6	3148.5	$N.m$	19.6
D2	3684.7/3199.4	3407.5	$N.m$	14.2
D3	3528.8/3292	3380.6	$N.m$	7
D4	3247.5/3149.3	3128	$N.m$	3
D5	3536.2/3402.8	3128	$N.m$	3.5
D6	3414.8/3353.5	3128	$N.m$	1.8

D6 are covered by iron, the design is made only with small saliency i.e. field axis or direct axis and inter polar or quadrature axis shows similar inductance. Thus the effect due to reluctance torque is almost negligible. Table 3.8 shows the peak to peak value of torque during load. The torque ripple for the surface mounted machines (D1 and D4) is large. The torque ripple for the pole-cap structure (D3 and D4) is less. If Fourier decomposition is made for the curves shown in Figure 3.27, the harmonics contents such as 6th and 12th may be present which is going to be discussed in the Chapter 4.

3.4.3. Relative Torque Pulsation Comparison

The comparisons of relative pulsating torque under load and no-load for all the geometries
are shown in Figure 3.28. The cogging torque and torque ripple are determined in percentage
according to the Equation 3.99. The comparison shows that the geometry D6 has less cogging
torque and torque ripple. D1 which is a surface mounted arrangement has higher percentage
of cogging torque and torque ripple. **It should be noted that the geometry which has
minimum cogging torque does also have minimum torque ripple.**

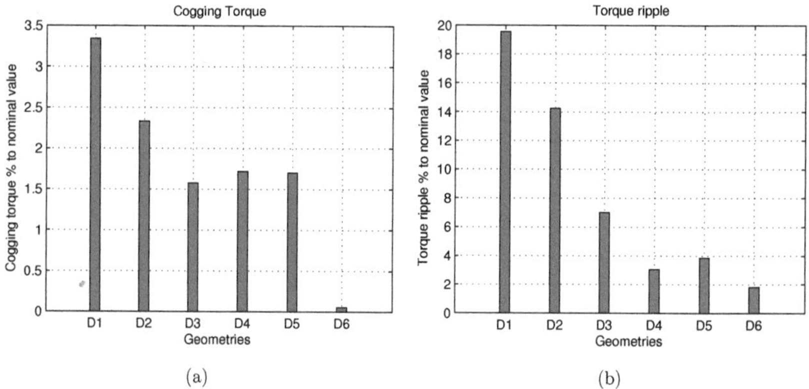

Figure 3.28.: Torque pulsation in [%]. (a) Without load. (b) With load.

3.4.4. THD from Surface Mounted to Pole-Cap

Three different rotor geometries have already been mentioned at the end of Chapter 2. The flux
density distribution of the surface mounted magnet and shell type magnet resemble square wave
distribution. It is desirable to have sinusoidal flux distribution at the air gap. This attempt is
made in the variation of the pole-cap structure. Field distribution for the three rotor geometries
are mathematically expressed as functions of sine and cosine. The field distribution harmonics

at no-load are compared with the factor total harmonic distortion (THD) for the three types
of rotor structures. The THD for the flux density is expressed as,

$$B_{THD} = \frac{1}{B_{\delta 1}} \sqrt{\sum_{\mu=2,3,\ldots}^{\infty} B_{\mu}^2} \qquad (3.100)$$

In Table 3.9, the calculated THD for geometries D1, D2, D3 and D6 are listed. In Table 3.9,
the geometries D3 and D6 with pole-cap structure have reduced THD factor.

Table 3.9.: Total Harmonic Distortion

Definition	THD %
D1	22.5
D2	24.4
D3	14.5
D6	14.5

4. Torque Analysis

The influence of permanent magnet motor geometrical characteristics on torque ripple has been widely studied and the references have been mentioned in Chapter 1. It is well known that the most influencing parameters for permanent magnet machine are slot opening and magnet pole coverage (ratio of pole-arc to pole-pitch). Therefore, these parameters are used for the sensitivity analysis for the considered geometry. In this chapter, the 2D harmonic analysis of cogging torque and torque ripple will be discussed. Recalling the torque mathematical formulation, "Torque is generated by the interaction of two sinusoidal fields in the air gap. The sinusoidal fields should have harmonics with identical pole pitch".

4.1. Pulsating Torque for Considered Geometries

Before analyzing the torque pulsation due to field harmonics, the relative torque pulsation (equation 3.99) under no-load and load for the geometries D1,D2,D4 and D5 are discussed in this section. Geometries D3 and D6 are not considered in this section due to geometry limitations. Slot opening and pole coverage are used for sensitivity analysis for the considered geometries. The calculations for torque pulsations are carried out numerically using FEM program FLUX from CEDRAT [60]. In this program, the analysis is performed as a transient magnetic application for a periodicity of one elementary machine. Three phase sinusoidal current is imposed for different rotor positions for constant speed.

The mesh at the air gap where the energy is concentrated, is very sensitive for the cogging torque analysis results FEMAG [61]. The numerical calculations may have poor convergence, if the geometry is poorly meshed at the air gap. In order to overcome this difficulty, the mesh structure is considered quadrilateral and divided evenly at the air gap i.e. for every rotation, the end nodes are placed exactly between the fixed part and the moving part (please refer Appendix D). For the analysis, the pole-coverage is selected between 0.53 and 0.9 due to geometry construction limitation. Further, the volume of the magnet for each geometry is maintained constant. The stator outer diameter and inner rotor diameter as well as cuurent densisty is kept constant for all the geometries.

4.1.1. Cogging Torque from D1 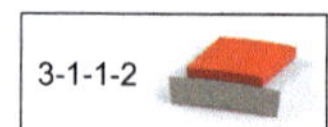

The relative cogging torque for different pole coverage and slot opening (SO) for D1 is shown in Figure 4.1(a). In the analysis, SO5 means slot opening width of $5mm$. Cogging torque at slot opening SO3 is less in comparison to SO5 and SO7. For SO5, the maximum cogging torque is 4% at pole coverage 0.6 and the minimum cogging torque is 1% at pole coverage . However, all the variations in slot opening width have minimum cogging torque at pole coverage range between $0.7 < \alpha < 0.73$. Therefore, under no-load operation the pole coverage value range between $0.7 < \alpha < 0.73$ is the optimal point. The cogging torque for different mechanical

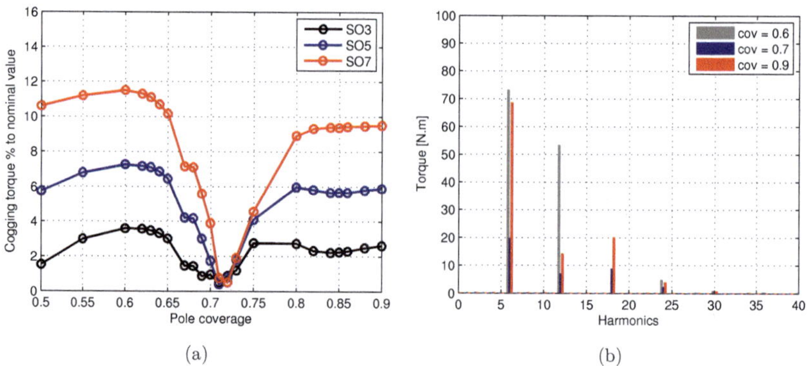

Figure 4.1.: Cogging Torque:(a)For various pole coverage and slot opening. (b) Torque harmonics @$5mm$.

position is already mentioned in section 3.4.1 and 3.4.2. If Fourier series is applied for the cogging torque, one can notice the presence of torque harmonic contents. For example, the spectral harmonics of cogging torque with SO5 for the geometry D1 is shown in Figure 4.1(b). As explained in the previous Chapter, the order of harmonic contents should be multiplied by the number of pole-pairs to obtain the ordinal numbers for the complete machine. The dominating harmonic components are the 6th and the 12th.

4.1.2. Cogging Torque from D2 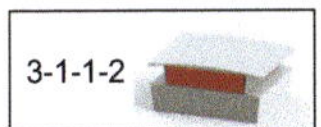

D2 has a shell type rotor arrangement as shown in Figure 2.17 (b). The relative cogging torque for different pole coverage and slot opening (SO) for D2 is shown in Figure 4.2(a). From the figure, it is evident that D2 has similar characteristics of cogging torque as D1. However, there is slight reduction in the relative torque pulsation for the pole coverage 0.7. Similar to D1, the cogging torque is minimum at range between $0.7 < \alpha < 0.73$ pole coverage for all the three slot opening widths(SO3, SO5 and SO7).

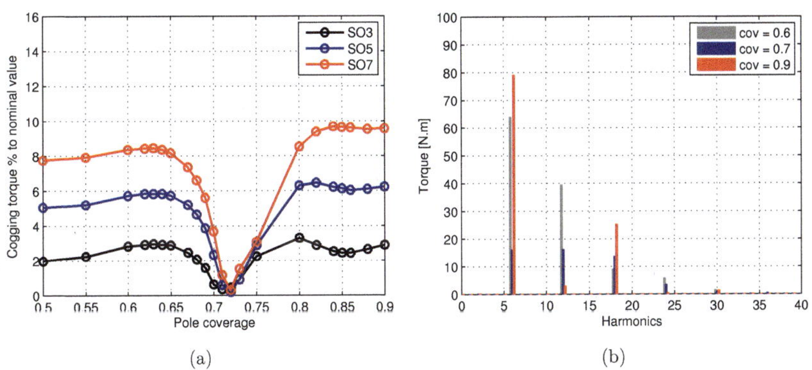

(a)   (b)

Figure 4.2.: Cogging Torque.(a)For various pole coverage and slot opening. (b) Torque harmonics @$5mm$.

The spectral harmonics content are shown for SO5 as an example in Figure 4.2(b). When observing the harmonic contents of cogging torque, the dominant ordinal numbers are again the 6th and the 12th. The 6th harmonic component is low for pole coverage 0.7 which is similar to D1.

4.1.3. Cogging Torque from D4 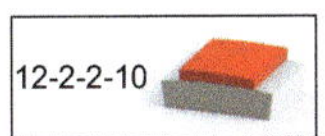

The relative cogging torque for different pole coverage and slot opening (SO) for D4 geometry can be seen in Figure 4.3(a). D4 has reduced cogging torque than D1 and D2. The variation of the cogging torque is between 0.1% and 4.2% for the considered slot opening widths. The point of minimum cogging torque happens at three ranges of pole coverages, between $0.53< \alpha <0.55$, between $0.7< \alpha <0.73$ and between $0.87< \alpha <0.9$. This is not the case for D1 and D2. The

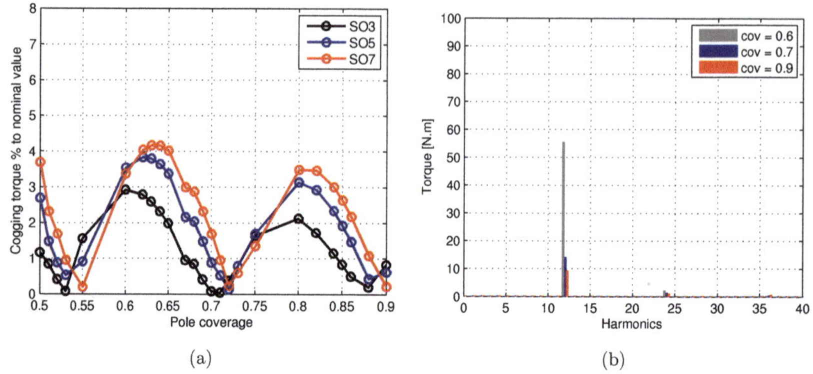

(a)   (b)

Figure 4.3.: Cogging Torque.(a)For various pole coverage and slot opening. (b) Torque harmonics @$5mm$.

cogging torque harmonic components present in D4 for slot opening of $5mm$ is shown in Figure 4.3(b)as an example. The dominating component is the 12th harmonics. The whole geometry for D4 has three elementary machines. The ordinal numbers shown in Figure 4.3(b) is for one-elementary machine.

4.1.4. Cogging Torque from D5

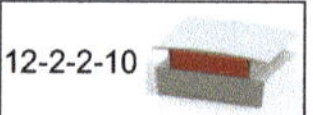

The relative cogging torque for different pole coverage and slot opening (SO) for D5 is shown in Figure 4.4(a). The point of minimum cogging torque happens at three ranges of pole coverages, between $0.53< \alpha <0.55$, between $0.7< \alpha <0.73$ and between $0.87< \alpha <0.9$, which is similar to D4. The cogging torque varies from 0.1% and 3% for all the three slot opening widths. The relative cogging torque for SO5 and SO7 is almost same for all pole coverages. In Figure 4.4(b), the corresponding cogging torque harmonic spectrum for SO5 is depicted. Similar to D4, the dominating component is the 12th at pole coverage 0.65.

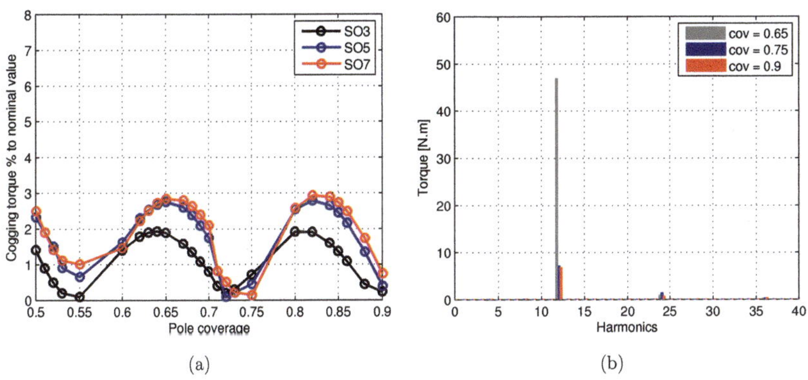

Figure 4.4.: Cogging Torque.(a)For various pole coverage and slot opening. (b) Torque harmonics @5mm.

4.1.5. Torque Ripple from D1 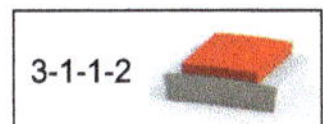

In this section, the relative torque ripple at rated load is discussed. Similar to no-load condition, the most influencing parameters are slot opening and pole coverage. The relative torque ripple for D1 for different pole coverage and slot opening width is shown in Figure 4.5(a). Contrary

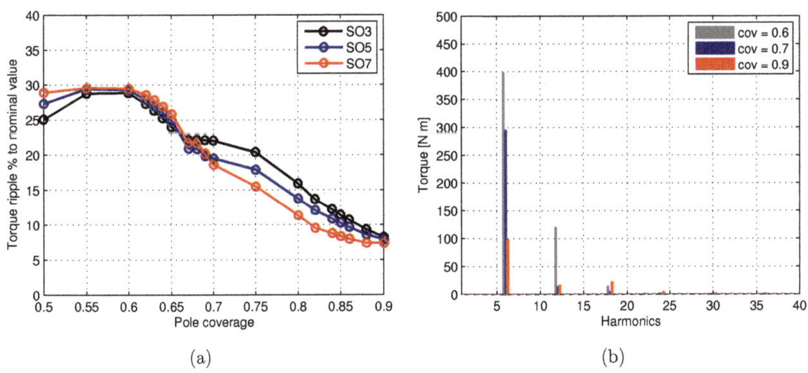

Figure 4.5.: Torque ripple.(a)For various pole coverage and slot opening. (b) Torque harmonics @5mm.

to the cogging torque for D1, the minimum torque ripple is shifted from pole coverage ranges between $0.7 < \alpha < 0.73$ to pole coverage ranges between $0.87 < \alpha < 0.9$. The reason for the shift is discussed in Chapter 4.3. The torque ripple varies between 7% and 30% for all the three slot opening widths (SO3, SO5 and SO7). In Figure 4.5(b), the torque harmonic components are shown for slot opening width 5mm.

4.1.6. Torque Ripple from D2 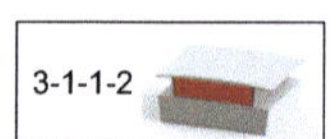

The relative torque ripple for D2 under rated load has similar characteristics as D1 under rated load. As shown in Figure 4.6(a), the minimum torque ripple shifts from pole coverage ranges between $0.7 < \alpha < 0.73$ to pole coverage ranges between $0.87 < \alpha < 0.9$ for all the three slot opening widths.The percentage of torque ripple varies only from 7% to 17% for all three slot opening (SO3, SO5 and SO7). The maximum torque ripple is reduced in comparison to D1 under load. In Figure 4.6(b), the torque harmonics are shown for slot opening width 5mm.

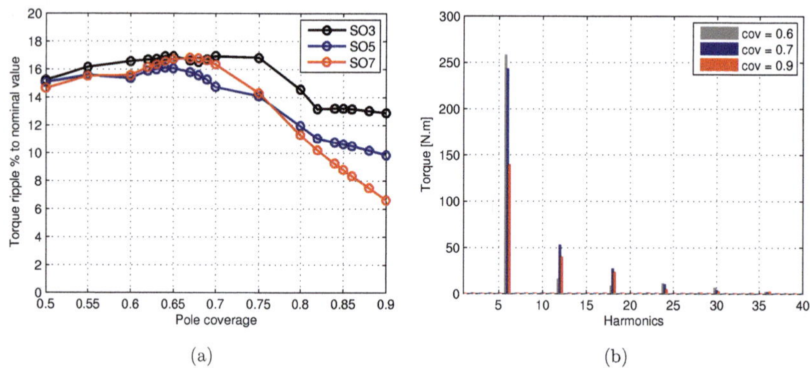

(a) (b)

Figure 4.6.: Torque ripple.(a)For various pole coverage and slot opening. (b) Torque harmonics @$5mm$.

4.1.7. Torque Ripple from D4 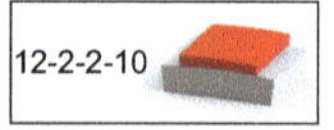

The relative torque ripple for D4 is different, when compared to D1 and D2 as shown in Figure 4.7(a). The torque ripple is minimum at three different pole coverages. This is similar to the no-load characteristics. However, the point of minimum torque is shifted. The relative torque ripple is in the range of 2% to 6.5%. At rated load, 6th and 18th harmonic components are present in addition to 12th and 24th harmonic components, which are the only components present during no-load.

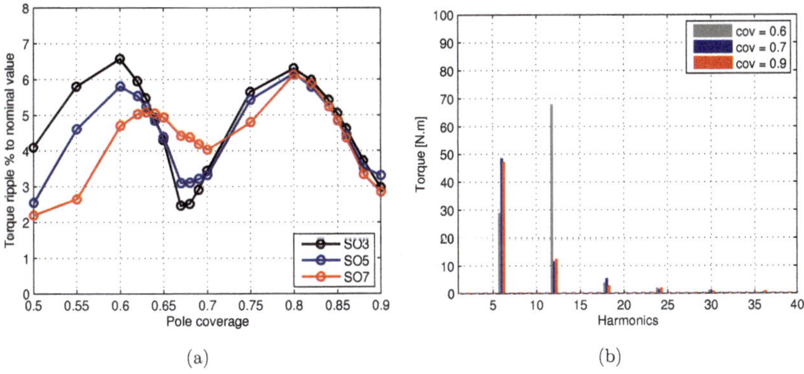

Figure 4.7.: Torque ripple.(a)For various pole coverage and slot opening. (b) Torque harmonics @$5mm$.

4.1.8. Torque Ripple from D5 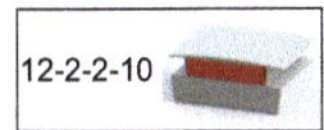

The relative torque ripple for D5 is shown in Figure 4.8(a). The characteristics of D5 is similar to D1 and D2. The relative torque ripple is in the range of 3% to 7%. The presence of ordinal numbers is shown in Figure 4.8(b) and it contains 2,4,6,12,18 and 24.

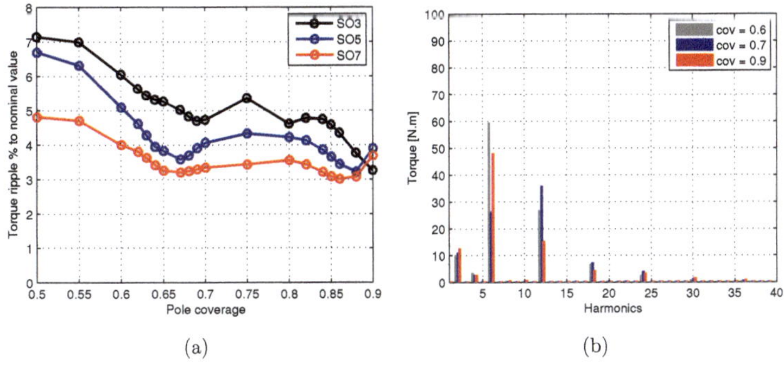

Figure 4.8.: Torque ripple.(a)For various pole coverage and slot opening. (b) Torque harmonics @$5mm$.

4.2. 2D Harmonics Analysis of Numerical Calculations

The detailed pulsating torque analysis of the considered geometries is explained in this section. As explained in the beginning of Chapter 3, the harmonics sources give clear picture about the design considerations.

Based on the permeance function, a theoretical approach for pulsating torque has been derived in the previous chapter for surface mounted PM machines. However, there is limitation with this approach for geometries having saturation in the regions such as iron ring and pole-cap. Nevertheless, this secondary effect problem can be easily handled in the numerical calculations. The approach used to analyze the pulsating torque in this work is 2D Fourier transformation of the air gap fields. The representation of the waves at the air gap comprises all the effects such as slot opening, saturation, etc.

The steps involved in 2D harmonics analysis are as follows:

4.2.1. Step 1

The initial step in the 2D Harmonic analysis [62], [64] and [65] method involves the use of the flux density distribution in the air gap, which is obtained from FEM calculations as a periodic waves both in space and time. These periodic wave functions can be represented as double Fourier series and it is given by,

$$f(p\theta, \omega t) = \int\limits_{n=-\infty}^{\infty} \int\limits_{m=-\infty}^{\infty} B(p\theta, \omega t) \cdot e^{-j(np\theta + m\omega t)} \tag{4.1}$$

where B is either radial or tangential flux density at the air gap B_r, B_θ over one elementary machine electrical period.

4.2.2. Step 2

Step 2 is converting the periodic flux density waves obtained from Step 1 into 2D Discrete Fast Fourier Transform (2DFFT). The obtained flux density from Step 1 is in the form of a matrix. The matrix A contains the distribution of flux density at the air gap for each rotor position. One elementary machine periodicity is considered for complete rotation i.e. the complete rotation is divided into equidistant steps of each rotor instant. It is expressed as

$$A = \begin{bmatrix} a_{11} & \cdots & a_{1N} \\ \vdots & \ddots & \vdots \\ a_{M1} & \cdots & a_{MN} \end{bmatrix} \tag{4.2}$$

This matrix can be transformed into 2DFFT using software tool MATLAB. It is computed using Equation 4.3,

$$G(m,n) = \sum_{i=0}^{M-1} \sum_{k=0}^{N-1} g(i,k) e^{-2\pi j(m\frac{i}{M}+n\frac{k}{N})} \tag{4.3}$$

This 2DFFT matrix Equation 4.3 is in complex form. It is further transformed to obtain the double Fourier coefficients and frequencies using,

$$S(m,n) = \frac{G(m,n)}{MN} \tag{4.4}$$

At this stage, the amplitude and phase of the flux density components (both tangential and radial) are available in space and time. As an example, Figure 4.9 shows 2D Fourier spectra of radial and tangential components. For the tangential component, the ordinal numbers are

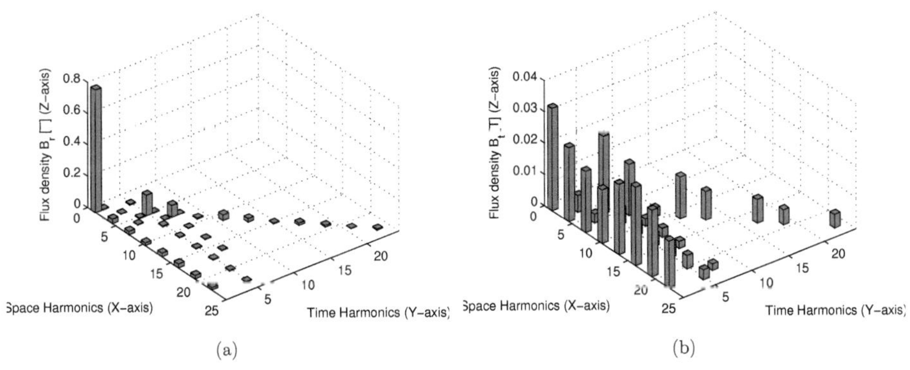

Figure 4.9.: Cogging Torque Matrix.(a)Radial Component. (b) Tangential Component

similar to radial component except for amplitude and phase angle. The harmonics that lie along the diagonal are odd harmonics and these are from permanent magnets which rotate along with rotor. The harmonics that are due to slot openings are along the X-axis in the plot. The dominant harmonics magnitude are conspicuous in the first time which is shown in Figure 4.9.

4.2.3. Step 3

As the next step, the matrix which has the information about amplitude and phase of flux density for different space and time can be represented as sinusoidal wave. This flux density wave is expressed as,

$$b(mp\theta, n\omega t) = \left|\frac{S(m,n)}{2}\right| \cos(mp\theta - n\omega t + \varphi) \tag{4.5}$$

where φ is the phase angle. Figure 4.10 shows two periodic waves of same space and different time for two pole period of 3-1-1-2 fractional slot winding arrangement. For example, one of

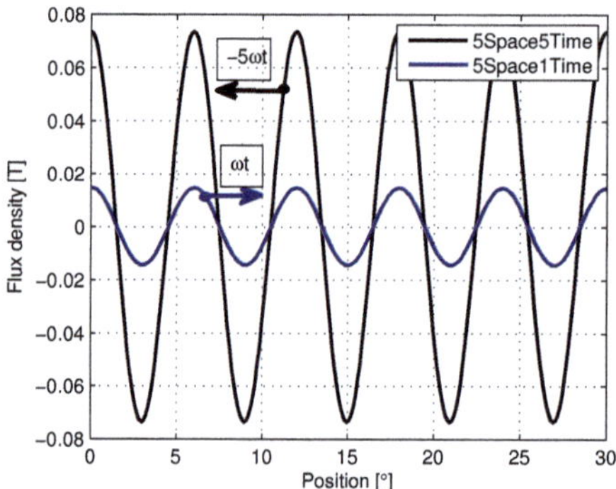

Figure 4.10.: Two periodic waves with different time.

the waves can be considered as radial component and the other can be considered as tangential component. Similarly, different dominant waves are selected from the matrix of arrays in step 2 process.

Harmonic pairs combinations are expressed in the form of short notations. These notations are used in this chapter in tables and definitions to point out the radial and tangential components. The notations meant are as follows,

- r-S5T1 meant radial component in 5th space and 1st time. This can be considered as first time elements.

- t-S55T11 meant tangential component in 55th space and 11th time. This can be considered as diagonal elements in the matrix.

4.2.4. Step 4

In the final step, the torque generating sinusoidal fields (radial and tangential components) in the air gap are separated out. In 2D harmonic analysis, the cogging torque and torque ripple are obtained using Maxwell Stress Tensor Equation 3.79 and it is rewritten here,

$$M = \frac{1}{\mu_0} r^2 \cdot l \int_0^{2\pi} B_r B_\theta \cdot d\theta \qquad (4.6)$$

where B_r is the radial and B_θ is the tangential flux density component at the air gap over one elementary machine electrical period.

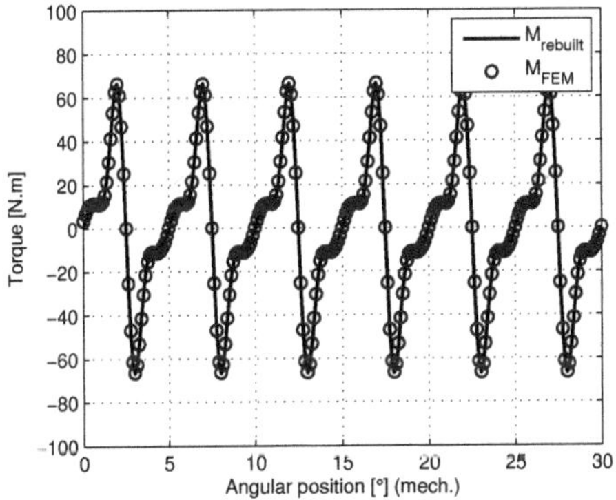

Figure 4.11.: Torque comparison.

Similar 2D Fourier analysis approach has been discussed in publications from [62], [63], [64], [65] and [66]. In publications [64] and [65] only the torque pulsation is studied under no-load. Moreover torque curve is not reconstructed from the harmonics, only an estimation of responsible harmonics is discussed. In another publication [62], similar reconstruction of cogging torque from harmonics is explained. However in this analysis, there is mismatch in the amplitude which could be a minor difference in the selection of harmonics for the generation of torque. But the tables discussed in publication [62] for the selected pairs for torque generation is of different space and time which is not according to the rule discussed in the chapter 3.3.1. It means that some pairs is not supposed to generate torque pulsation. In publications from [63] and [66], the selection of harmonic pairs is based on force density pairs and it is mentioned that reconstruction of torque directly using flux densities at the air gap has complexity. But in this work, torque is reconstructed using flux density pairs at the air gap. As an example, Figure 4.11 shows the torque directly obtained from FEM (blue circle), which is based on energy method [60] and the reconstruction of the torque from harmonics (black curve) based on Equation 4.6 under no-load operation of 3-1-1-2 fractional slot winding arrangement. For one elementary machine, as expected, there are six periods of torque pulsation. The calculated torque from the energy method and the reconstructed torque using 2D Fourier analysis are in good agreement.

In the following sections, the harmonics combination which are necessary to construct the torque for all the considered geometries are discussed. Due to the vast amount of data which cannot be covered in the work, only one slot opening width and three different pole coverage are studied for the considered geometries. The reason for selection of three pole coverages for the detailed analysis is because, the pole coverage in x-axis in Figure 4.1a for D1 geometry are divided in to three regions: the region which has minimum range (between $0.7 < \alpha < 0.73$), the region left to the minimum range (between $0.5 < \alpha < 0.7$) and region right to the minimum range (between $0.73 < \alpha < 0.9$). Similar case is also considered for other type of geometries. The geometries studied for pulsating torque analysis are

- D1 geometry for pole coverage (0.6, 0.7 and 0.9) in both no-load and load operation.

- D2 geometry for pole coverage (0.6, 0.7 and 0.9) in both no-load and load operation.

- D3 geometry at pole coverage 0.7 in both no-load and load operation.

- D4 geometry for pole coverage (0.6, 0.7/0.75) in no-load and load operation.

- D5 geometry for pole coverage (0.55/0.65, 0.7/0.75) in no-load and load operation.

- D6 geometry at pole coverage 0.7 in both no-load and load operation.

Table 4.1.: Example Diagonal and Time Harmonics DX pole coverage 0.6, SO $5mm$.

Diagonal	Br	Br Phase	Bt	Bt Phase	S1 Time	Br	Br Phase	Bt	Bt Phase
S55T11	0.0197	90	0.0088	-90	S55T1	0.0241	0	0.0241	180

The selected harmonics amplitude and phase angle are discussed using Figures and Tables. For example, the parameter name Diagonal in the tabular column (Table 4.1) represents the harmonics that rotate with rotor. The parameter name S1 Time in the tabular column (Table 4.1) represents the space harmonics of first time. These harmonics shown in the Table are the required ones to reconstruct the torque as explained in Step 4. Moreover, polar plot is used to show a qualitative impression of flux density in the polar grid. In the polar grid plane in order to show the higher order harmonics, the value of fundamental harmonic which is pointed as shown in Figure 4.12 is not displayed in the detailed analysis. In the polar grid plane, the negative phase angle $-90°$ is not drawn in clockwise and it is drawn as phase angle $270°$ in the counterclockwise.

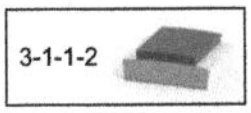

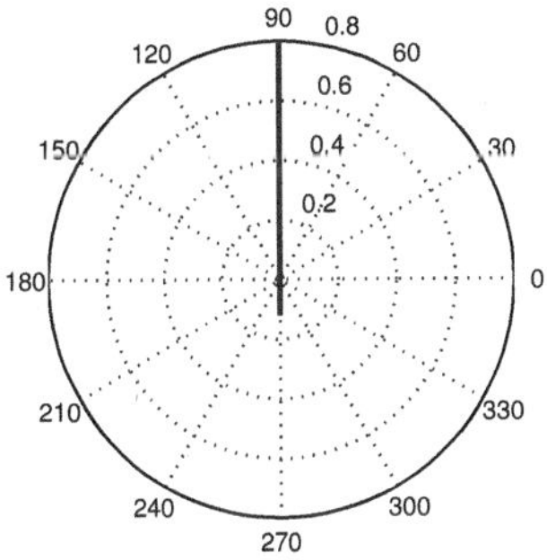

Figure 4.12.: Example: Polar plot radial diagonal elements.

4.3. Torque Analysis for D1

It can be understood from the beginning of this Chapter that for D1, the dominant torque harmonic components are the 6th and 12th. The idea behind the 2D Fourier analysis is to select the harmonics magnitude and phase angle from the radial and tangential components of flux density, which generate significant parasitic torque of 6th and 12th.

No-load analysis for pole coverage at 0.7:

Figure 4.13 shows the polar plot of flux density for radial and tangential components for pole coverage 0.7. This pole coverage has minimum relative torque. To depict other harmonics more clearly in the polar plot, the fundamental component is not displayed. The diagonal elements in Figure 4.13 (a),(b) are rotating along with rotor both in space and time. The dominant space

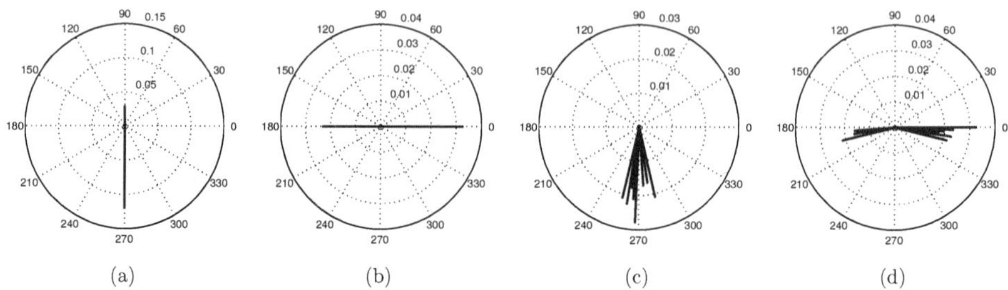

Figure 4.13.: Polar plot D1 pole coverage 0.7, SO $5mm$, $I = 0A$. (a) Diagonal Elements Radial. (b) Diagonal Elements Tangential (c) First Time Elements Radial. (d) First Time Elements Tangential.

harmonics are available in the first time (please refer Figure 4.9). These harmonics generate the expected pulsating torque under no-load. Table 4.2 lists the diagonal and time components for magnitude and phase angle of no-load flux density. At 0.7 pole coverage, the torque pulsation is due to the pairs $r - S5T5$, $t - S5T1$ and $r - S7T7$, $t - S7T1$. The phase angle of the diagonal elements that rotate both in space and time are either $+90°$ or $-90°$.

Table 4.2.: Diagonal and Time Harmonics D1 pole coverage 0.7, SO $5mm$.

Diagonal	Br	Br Phase	Bt	Bt Phase	1S Time	Br	Br Phase	Bt	Bt Phase
S1T1	0.7911	89.99	0.03279	0.0034	S4T1	0.0278	-92.25	0.02345	-2.244
S5T5	0.1195	-90.02	0.02291	179.9	S5T1	0.0107	-85.5	0.0159	-175
S7T1	0.08582	-90.02	0.01885	179.9	S7T1	0.0209	-94.5	0.0198	-4.507
S11T11	0.0301	89.96	0.0108	-0.0344	S11T1	0.0163	-80	0.0161	-170.9
S13T13	0.02838	89.9	0.01131	-0.04088	S13T1	0.0226	-99	0.0228	-9
S17T17	0.01021	-90	0.0053	180	S17T1	0.0212	-76.4	0.0212	-166.4
S19T19	0.01188	-90	0.0066	180	S19T1	0.0211	-103.6	0.0214	-13.6

No-load analysis for pole coverage at 0.6:

The maximum relative torque for the geometry D1 is at pole coverage 0.6. The corresponding polar plot is shown in Figure 4.14. The phase angle for diagonal elements are similar to pole coverage 0.7. The no-load values for pole coverage 0.6 are tabulated in Table 4.3. The possible 6th harmonic in the pulsating torque is from $r - S5T5$ and $t - S5T1$ radial and tangential components. The 7th radial component is less when compared to pole coverage 0.7. The possible 12th component is from $r - S11T11$ and $t - S11T1$ radial and tangential components.

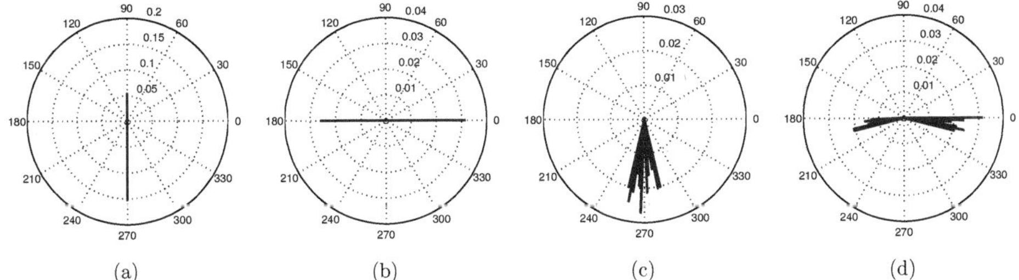

Figure 4.14.: Polar plot pole coverage 0.6, SO $5mm$, $I = 0A$. (a) Diagonal Elements Radial. (b) Diagonal Elements Tangential (c) First Time Elements Radial. (d) First Time Elements Tangential.

Table 4.3.: Diagonal and Time Harmonics D1 pole coverage 0.6, SO $5mm$.

Diagonal	Br	Br Phase	Bt	Bt Phase	1S Time	Br	Br Phase	Dt	Dt Phase
S1T1	0.7446	90	0.0316	0.0032	S2T1	0.0212	-87.8	0.013	-177.8
S3T3	0.1054	-90	0.0182	180	S4T1	0.0271	-92.25	0.0238	-2.245
S5T5	0.1515	-90	0.0257	180	S5T1	0.0162	-85.5	0.0157	-175.5
S7T7	0.0137	-90	0.0002	180	S7T1	0.0193	-94.5	0.0188	-4.505
S9T9	0.05577	90	0.0165	-0.0278	S10T1	0.0157	-96.8	0.0157	-6.8
S11T11	0.0253	90	0.0008	-0.0354	S11T1	0.0146	-90	0.0144	-170.9
S13T13	0.018	-90	0.0076	180	S13T1	0.02126	-99	0.0211	-9
S17T17	0.0023	-90	0.0014	-0.05	S17T1	0.0202	-76.5	0.0204	-166.4
S19T19	0.0119	90	0.0066	-0.06	S19T1	0.0201	-103.6	0.026	-13.6

No-load analysis for pole coverage at 0.9:

The polar plot for pole coverage 0.9 is shown in Figure 4.15. There is a phase change in $5^{th}, 7^{th}, 11^{th}$ and 13th Diagonal radial components when compared to the pole coverage 0.7. There is no change in $5^{th}, 7^{th}, 11^{th}$ and 13th components for the first time elements.

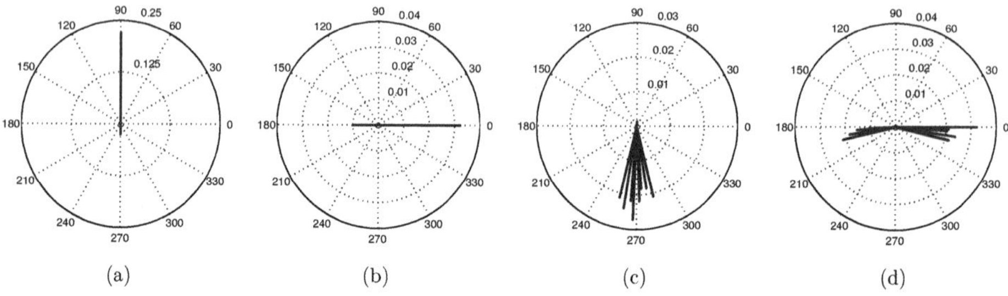

Figure 4.15.: Polar plot D1 pole coverage 0.9, SO $5mm$, $I = 0A$. (a) Diagonal Elements Radial. (b) Diagonal Elements Tangential (c) First Time Elements Radial. (d) First Time Elements Tangential.

The significant 6th torque harmonic is from flux density component 5th harmonic combination pairs as given in Table 4.4. The 12th torque harmonic component is contributed from the flux density component 13th harmonic combination pairs.

Table 4.4.: Diagonal and Time Harmonics D1 pole coverage 0.9, SO $5mm$.

Diagonal	Br	Br Phase	Bt	Bt Phase	S1 Time	Br	Br Phase	Bt	Bt Phase
S1T1	0.8218	90	0.0325	0.0037	S2T1	0.0218	-87.8	0.0104	-177.8
S3T3	0.1054	-90	0.0182	180	S3T1	0.0014	88.4	0.0001	-0.32
S5T5	0.0803	90	0.0117	-0.0148	S5T1	0.0182	-85.5	0.0153	-175.5
S7T7	0.0174	90	0.0023	-0.02	S7T1	0.0216	-94.5	0.02	-4.5
S11T11	0.0234	-90	0.0084	180	S11T1	0.0183	-81	0.0183	-171
S13T13	0.0247	-90	0.01	180	S13T1	0.024	-99	0.0239	-9
S17T17	0.0154	-90	0.0078	180	S17T1	0.021	-76.4	0.0213	-166.4
S19T19	0.0097	-90	0.0056	179.5	S19T1	0.0213	-103.6	0.0216	-13.6

<u>Load analysis for pole coverage at 0.7:</u>

Torque ripple under load current is from the contribution of harmonics from both stator and rotor. When looking at the current sheet distribution, fractional slot winding arrangement has different harmonics order particularly (Table 3.2). Contrary to the no-load operation,

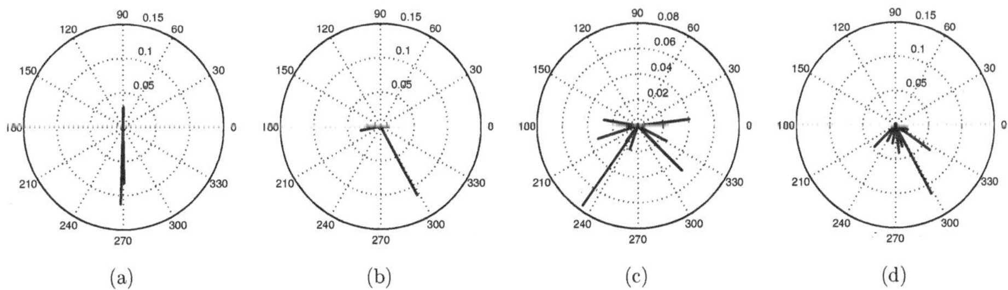

Figure 4.16.: Polar plot D1 pole coverage 0.7, SO $5mm$, $I = I_N$. (a) Diagonal Elements Radial. (b) Diagonal Elements Tangential (c) First Time Elements Radial. (d) First Time Elements Tangential.

torque pulsation appears from other harmonic orders. The flux density at load operation has the effect of both stator and rotor fields. Figure 4.16 shows the polar plot of D1 for pole coverage 0.7 under load. Due to the armature reaction, the field is disturbed when compared to the polar plot under no-load. Moreover, there are appearances of additional harmonic components. The significant harmonics amplitude contribution is from $r - S5T5$, $t - S5T1$ harmonics pairs. The values of all the harmonics order are provided in the Table 4.5. There is no change of phase angle of the 5th harmonics in the Diagonal elements but it has changed in the Time elements.

Table 4.5.: Diagonal and Time Harmonics D1 pole coverage 0.7, SO $5mm$.

Diagonal	Br	Br Phase	Bt	Bt Phase	1S Time	Br	Br Phase	Bt	Bt Phase
S1T1	0.789	114.1	0.1148	-61.6	S4T1	0.0757	-125.4	0.063	-34.8
S5T5	0.1122	-92.02	0.0298	-169.6	S5T1	0.0489	-45.2	0.0449	-135.3
S7T7	0.082	-89.2	0.0226	176.5	S7T1	0.0151	-127.9	0.014	-38.4
S11T11	0.0293	88.7	0.0116	6.268	S11T1	0.0406	7	0.0404	-82.7
S13T13	0.0276	90.8	0.0121	-2.16	S13T1	0.0327	-162.7	0.0328	-72.3
S17T17	0.01	-90	0.0055	-177	S17T1	0.0254	-27.6	0.0257	-117.7
S19T19	0.01159	-89.6	0.007	179.2	S19T1	0.019	-109.6	0.0192	-19.4

The influence of the dominant harmonics of $r - S5T5$ and $t - S5T1$ pairs is shown in the Figure 4.17(b). Figure 4.17 illustrates torque for two pole period mechanical. The left diagram 4.17(a) is with 5th order component where reconstruction of torque follows the torque obtained from FEM. The right diagram 4.17(b) is without 5th order component. It is clear that there is difference in the torque pulsation magnitude.

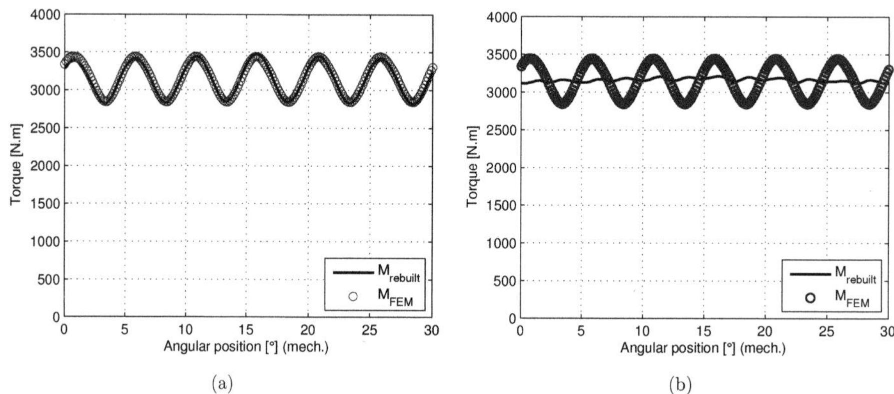

Figure 4.17.: Torque ripple D1 pole coverage 0.7, SO $5mm$, $I = I_N$. Torque ripple (a)Basic components. (b)Reduced components.

Load analysis for pole coverage at 0.6:

According to Figure 4.5, torque ripple is higher under load at pole coverage 0.6. The polar plot for pole coverage 0.6 is shown in Figure 4.18. There are presence of other harmonic orders 3rd and 9th. There is only slight change in the phase angle. Table 4.6 gives the values of torque producing harmonic orders for pole coverage 0.6 under load. The 3rd order and the 5th order contributes to the torque ripple. The relative torque ripple is increased when compared to the pole coverage 0.7 because the magnitude of the 5th harmonics is increased. Figure 4.19 shows the reconstruction of torque with and without 3rd and 5th harmonics order. It can be inferred that there is reduction in the torque ripple if there is no 3rd and 5th harmonic component, as depicted in Figure 4.19(b).

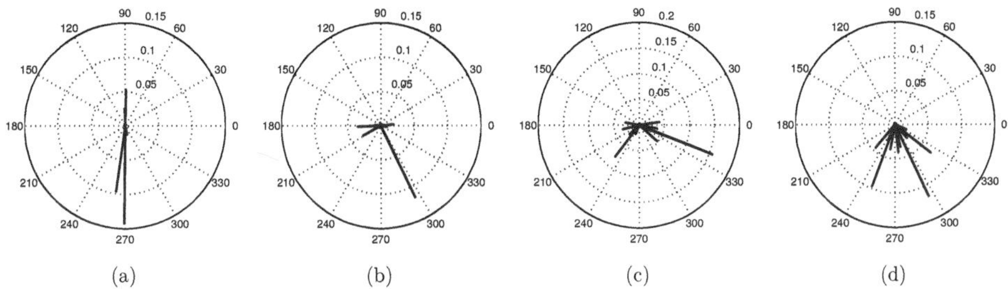

(a) (b) (c) (d)

Figure 4.18.: Polar plot D1 pole coverage 0.6, SO $5mm$, $I = I_N$. (a) Diagonal Elements Radial. (b) Diagonal Elements Tangential (c) First Time Elements Radial. (d) First Time Elements Tangential.

Table 4.6.: Diagonal and Time Harmonics D1 pole coverage 0.6, SO $5mm$.

Diagonal	Br	Br Phase	Bt	Bt Phase	1S Time	Br	Br Phase	Bt	Bt Phase
S1T1	0.7488	113.4	0.117	-64	S2T1	0.1564	-21.4	0.097	-111.3
S3T3	0.096	-98.3	0.03	-151.3	S4T1	0.077	-127.9	0.067	-37.4
S5T5	0.142	-90.5	0.035	-177.5	S5T1	0.0471	-41	0.0447	-131.3
S7T7	0.0132	-77.7	0.004	105.5	S7T1	0.0173	-126.3	0.0166	-35.9
S9T9	0.054	88.7	0.02	5.2	S10T1	0.02	169.7	0.027	-100.3
S11T11	0.0245	92.3	0.009	-8.8	S11T1	0.0406	7.4	0.041	-82.4
S13T13	0.0176	-91.5	0.008	-173.6	S13T1	0.033	-165.4	0.033	-75.4
S17T17	0.0024	84.5	0.0014	21.1	S17T1	0.025	-25.1	0.0252	-115
S19T19	0.012	90.5	0.007	-0.37	S19T1	0.0182	-112	0.0186	-22.1

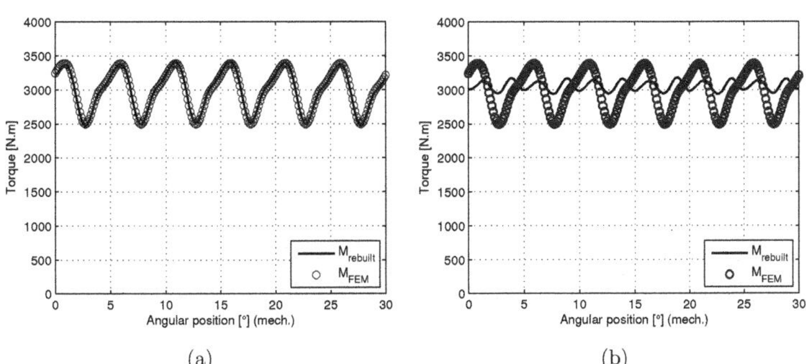

(a) (b)

Figure 4.19.: Torque Ripple D1 pole coverage 0.6, SO $5mm$, $I = I_N$. Torque ripple (a)Basic components. (b)Reduced components.

<u>Load analysis for pole coverage at 0.9:</u>

The lowest relative torque ripple is at pole coverage 0.9 for D1 geometry under load. The phase angle of 3rd, 5th, 7th, 11th and 17th is changed for the Diagonal elements. Figure 4.20 shows the polar plot of D1 at pole coverage 0.9.

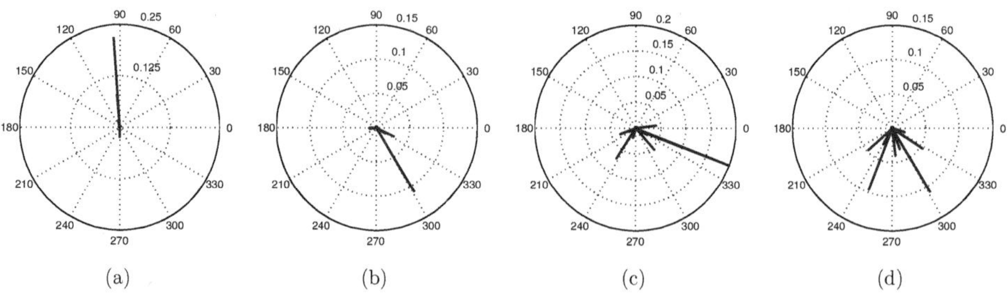

(a)   (b)   (c)   (d)

Figure 4.20.: Polar plot D1 pole coverage 0.9, SO $5mm$, $I = I_N$. (a) Diagonal Elements Radial. (b) Diagonal Elements Tangential (c) First Time Elements Radial. (d) First Time Elements Tangential.

Table 4.7 shows that the 3rd harmonics is higher when compared to pole coverages 0.6 and 0.7. However, 5th and 17th harmonics are minimum. The working harmonics for the torque pulsation are $r - S5T5$, $t - S5T1$ and $r - S7T7$, $t - S7T1$ radial and tangential components. For this pole coverage, the pairs $r - S3T3$ and $t - S3T3$ contribute to the DC shift of the torque

Table 4.7.: Diagonal and Time Harmonics D1 pole coverage 0.9, SO $5mm$.

Diagonal	Br	Br Phase	Bt	Bt Phase	1S Time	Br	Br Phase	Bt	Bt Phase
S1T1	0.8125	116.5	0.12	-58.7	S2T1	0.2	-21.2	0.0957	-111
S3T3	0.217	94.2	0.03	-25.3	S3T1	0.0013	92.3	0.0002	-21.4
S5T5	0.0784	95.1	0.016	-30.8	S5T1	0.0556	-48.3	0.0473	-138
S7T7	0.0169	103.6	0.007	-59.5	S7T1	0.0122	-156.2	0.011	-65.5
S11T11	0.0224	-90.8	0.0094	-172.83	S11T1	0.0403	7.1	0.04	-83
S13T13	0.0238	-90	0.011	-177.5	S13T1	0.0322	-159.8	0.032	-69.9
S17T17	0.0152	-89.6	0.008	178.6	S17T1	0.0206	-29.2	0.0264	-119.4
S19T19	0.0096	-89.5	0.005	177.2	S19T1	0.0186	-106.5	0.019	-16.4

component from the average torque. Figure 4.21 shows that the characteristics of the curve from FEM and the reconstructed curve with the contribution of all the harmonics are similar except to the DC shift caused by the 3rd order. The relative torque ripple is less due to the reduced magnitude of other harmonics when compared to pole coverage 0.6 and 0.7.

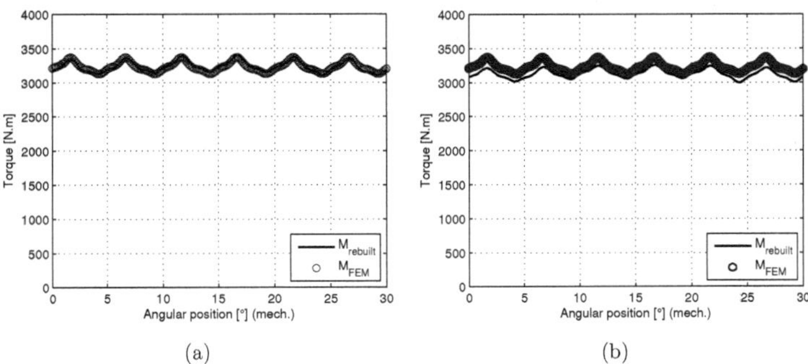

(a)

(b)

Figure 4.21.: Torque ripple D1 pole coverage 0.9, SO $5mm$, $I = I_N$. (a)Basic components. (b)Without 3rd.

4.4. Torque Analysis for D2

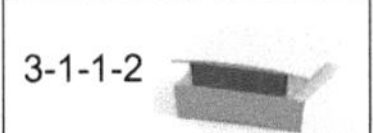

No-load analysis for pole coverage at 0.7:

The relative cogging torque characteristics for D2 are similar to D1 as shown in Figure 4.2 and 4.6. As discussed earlier, the torque harmonics are of the order 6th, 12th and 18th. The polar plot of D2 at pole coverage 0.7 in Figure 4.22 (a) shows that the phase angle have either $+90°$ or $-90°$ for the Diagonal radial components. The harmonics that generates the expected pulsating torque under no-load are $r - S5T5$, $t - S5T1$ and $r - S7T7$, $t - S7T1$. Table 4.8 lists

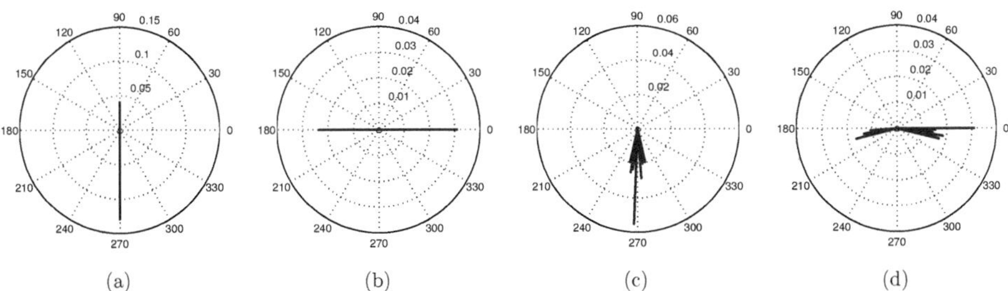

(a)

(b)

(c)

(d)

Figure 4.22.: Polar plot D2 pole coverage 0.7, SO $5mm$, $I = 0A$. (a) Diagonal Elements Radial. (b) Diagonal Elements Tangential (c) First Time Elements Radial. (d) First Time Elements Tangential.

the magnitude and phase angle for diagonal and time components. The magnitude and phase angle are similar to D1. Hence, the relative cogging torque value is same as D1. The other harmonics pairs such as $r - S11T11$, $t - S11T1$, $r - S13T13$, $t - S13T1$, $r - S17T17$, $t - S17T1$ and $r - S19T19$, $t - S19T1$ generates 12th and 18th torque harmonics with low magnitude.

Table 4.8.: Diagonal and Time Harmonics D2 pole coverage 0.7, SO $5mm$.

Diagonal	Br	Br Phase	Bt	Bt Phase	1S Time	Br	Br Phase	Bt	Bt Phase
S1T1	0.795	90	0.0312	0.003	S4T1	0.0553	-92.3	0.0154	-2.24
S5T5	0.1295	-90	0.0234	180	S5T1	0.0286	-85.5	0.0105	-175.5
S7T7	0.0791	-90	0.0174	180	S7T1	0.0316	-94.5	0.0142	-4.507
S11T11	0.0416	89.9	0.0145	-0.034	S11T1	0.0201	-80.9	0.0133	-170.9
S13T13	0.0267	89.9	0.0107	-0.04	S13T1	0.0252	-99	0.0186	-9.1
S17T17	0.0162	-90	0.0082	180	S17T1	0.0196	-76.5	0.0168	-166.4
S19T19	0.0114	-90	0.0063	180	S19T1	0.0196	-103.6	0.0171	-13.6

No-load analysis for pole coverage at 0.6:

The polar plot of relative cogging torque at pole coverage 0.6 is shown in Figure 4.23. The phase angle of the harmonics 7th, 13th, 17th and 19th is changed for pole coverage 0.6.

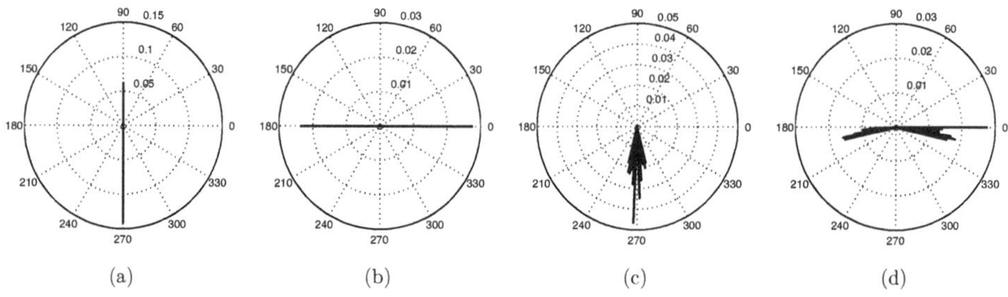

Figure 4.23.: Polar plot D2 pole coverage 0.6, SO $5mm$, $I = 0A$. (a) Diagonal Elements Radial. (b) Diagonal Elements Tangential (c) First Time Elements Radial. (d) First Time Elements Tangential.

The harmonics values are tabulated in Table 4.9. The 5th order is increased and the pairs $r - S5T5$, $t - S5T1$ is the most dominating torque generating component. The other harmonics 7th, 11th, 13th, 17th and 19th also contribute to the torque pulsation. There is also presence of even harmonics at this pole coverage. It does not contribute to torque, since its torque producing harmonic pairs are not present.

Table 4.9.: Diagonal and Time Harmonics D2 pole coverage 0.6, SO $5mm$.

Diagonal	Br	Br Phase	Bt	Bt Phase	1S Time	Br	Br Phase	Bt	Bt Phase
S1T1	0.7134	90	0.0277	0.0031	S4T1	0.0465	-92.25	0.0139	-2.245
S5T5	0.1419	-90	0.0234	180	S5T1	0.026	-85.5	0.001	-175.5
S7T7	0.0106	90	0.0041	-0.02	S7T1	0.0278	-94.5	0.013	-4.505
S9T9	0.063	90	0.0181	-0.0277	S10T1	0.0209	-96.8	0.0131	-6.789
S11T11	0.0107	90	0.0032	-0.0352	S11T1	0.0183	-81	0.0124	-171
S13T13	0.029	-90	0.0117	180	S13T1	0.0228	-99	0.0168	-9
S17T17	0.011	90	0.0056	-0.053	S17T1	0.018	-76.4	0.0154	-166.5
S19T19	0.0118	90	0.0066	-0.06	S19T1	0.0177	-103.6	0.0157	-13.6

No-load analysis for pole coverage at 0.9:

D2 also have another high relative cogging torque at pole coverage 0.9. The Diagonal elements such as 5th, 11th and 13th have their phase angle related to pole coverage of 0.7 changed. It is shown in the polar plot Figure 4.24. As listed in the Table 4.10, there are presence of

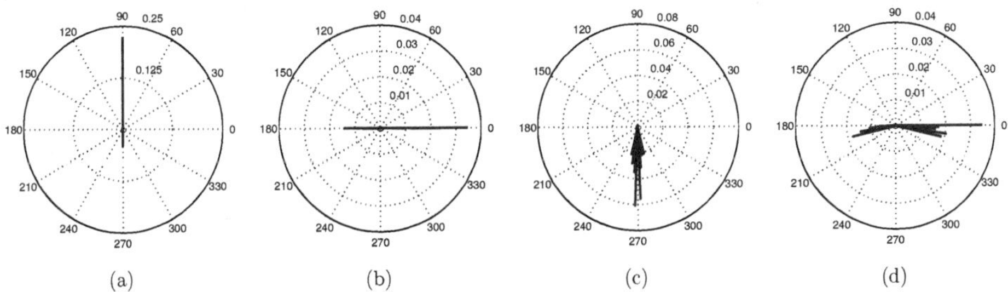

(a)　　　　(b)　　　　(c)　　　　(d)

Figure 4.24.: Polar plot D2 pole coverage 0.9, SO $5mm$, $I = 0A$. (a) Diagonal Elements Radial. (b) Diagonal Elements Tangential (c) First Time Elements Radial. (d) First Time Elements Tangential.

3rd harmonics in this pole coverage. 5th and 7th diagonal elements are reduced. Yet, the time elements of 5th and 7th are increased when compared to the pole coverage of 0.7. In this case, the major torque contributing component is the combination of harmonic pairs $r - S5T5$ and $t - S5T1$.

Table 4.10.: Diagonal and Time Harmonics pole coverage 0.9, SO 5mm.

Diagonal	Br	Br Phase	Bt	Bt Phase	1S Time	Br	Br Phase	Bt	Bt Phase
S1T1	0.8728	90	0.0347	0.0028	S2T1	0.0554	-87.8	0.0074	-178
S3T3	0.2238	-90	0.0236	180	S3T1	0.0014	88.4	0.0001	-0.6
S5T5	0.0655	90	0.009	-0.0147	S5T1	0.0312	-85.5	0.011	-175.5
S7T7	0.0055	-90	0.0032	180	S7T1	0.0349	-94.5	0.016	-4.5
S11T11	0.0403	-90	0.0142	180	S11T1	0.0217	-81	0.014	-171
S13T13	0.034	-90	0.014	180	S13T1	0.0283	-99	0.0203	-9
S17T17	0.0128	-90	0.0065	180	S17T1	0.021	-76.4	0.0179	-166.5
S19T19	0.0046	-90	0.0025	179.5	S19T1	0.0214	-103.6	0.0186	-13.6

<u>Load analysis for pole coverage at 0.7:</u>

Torque pulsation under load for D2 also has similar characteristics as D1 under load (please refer Figure 4.6). For the pole coverage 0.7, the average torque is higher than D1. This is due to the increase in fundamental component, which is not shown in the polar plot 4.25. From the polar plot Figure 4.25 the phase angle change of the harmonics due to the stator field can be seen. Table 4.11 lists the values of the harmonics for pole coverage 0.7. All the dominant

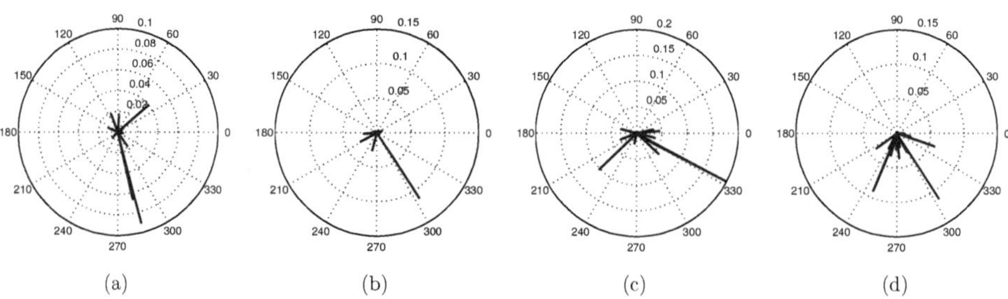

(a)   (b)   (c)   (d)

Figure 4.25.: Polar plot D2 pole coverage 0.7, SO 5mm, $I = I_N$. (a) Diagonal Elements Radial. (b) Diagonal Elements Tangential (c) First Time Elements Radial. (d) First Time Elements Tangential.

diagonal harmonics, which are of significant amplitude in D2 are less in comparison to D1. These harmonics are 5th, 7th, 11th, 13th, 17th and 19th. The highest torque pulsating amplitude is from the pairs $r - S5T5$ and $t - S5T1$. When pairs $r - S5T5$ and $t - S5T1$ component is assumed zero, the pulsation torque seems to be nearly constant. This is depicted in Figure 4.26 (b). When the harmonics content such as $r - S5T5$, $t - S5T1$, $r - S7T7$, $t - S7T1$, $r - S11T11$, $t - S11T1$, $r - S13T13$, $t - S13T1$, $r - S17T17$, $t - S17T1$ and $r - S19T19, t - S19T1$ are included, the curve of FEM and reconstructed curve are having good agreement.

Table 4.11.: Diagonal and Time Harmonics D2 pole coverage 0.7, SO 5*mm*.

Diagonal	Br	Br Phase	Bt	Bt Phase	1S Time	Br	Br Phase	Bt	Bt Phase
S1T1	0.8486	114.1	0.1138	-56.5	S4T1	0.101	-136.9	0.0583	-18.5
S5T5	0.09	-75.1	0.0263	-151.3	S5T1	0.0587	-42.2	0.0371	-144.7
S7T7	0.0664	-76.8	0.0186	-176.6	S7T1	0.0115	-158.5	0.001	1.18
S11T11	0.0169	85.5	0.006	13.4	S11T1	0.0448	5.7	0.0356	-84.5
S13T13	0.0187	112.3	0.0076	19.7	S13T1	0.0337	-158.8	0.0301	-61.3
S17T17	0.007	-134.6	0.0032	-135	S17T1	0.0253	-28	0.0237	-119.9
S19T19	0.005	-73.9	0.003	-168.1	S19T1	0.02	-102.7	0.02	-11.3

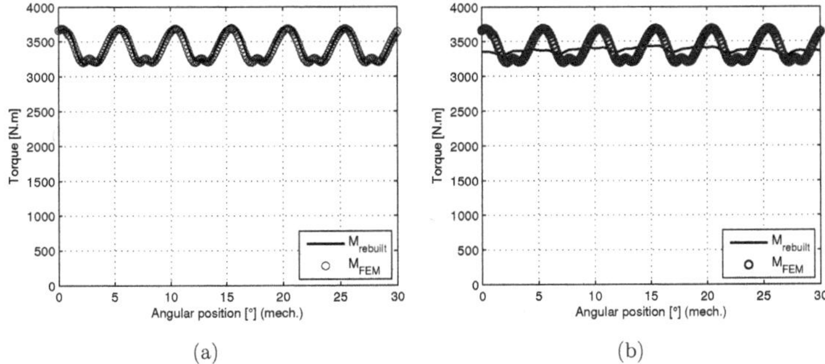

(a)   (b)

Figure 4.26.: Torque ripple D2 pole coverage 0.7, SO 5*mm*, $I = I_N$. (a)Basic components. (b)Reduced components.

Load analysis for pole coverage at 0.6:

The polar plot is shown in Figure 4.27 for pole coverage 0.6. There is appearance of 3rd harmonics and its phase angle is negative. Phasors of other harmonics are displaced under load for this pole coverage than under no-load. Table 4.12 shows the results of Diagonal elements and Time elements for pole coverage 0.6. The table results can be compared with the D1 configuration under load. The fundamental amplitude of radial component is higher in D2 than D1. The phase angle for the diagonal elements is negative and its value is different from D1. The phase angle value of the time elements is nearly same as D1.

The comparison of reconstructed torque curve with FEM curve (see Figure 4.28) shows that the torque pulsation can be reduced without 3rd and 5th order (refer Figure 4.28(b)). Moreover, 3rd

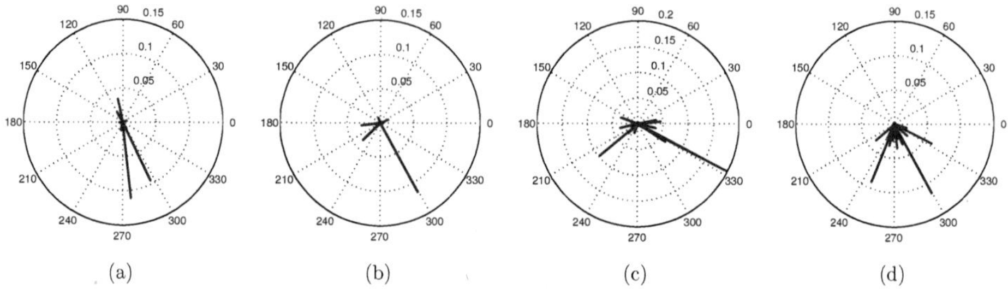

Figure 4.27.: Polar plot D2 pole coverage 0.6, SO $5mm$, $I = I_N$. (a) Diagonal Elements Radial. (b) Diagonal Elements Tangential (c) First Time Elements Radial. (d) First Time Elements Tangential.

Table 4.12.: Diagonal and Time Harmonics D2 pole coverage 0.6, SO $5mm$.

Diagonal	Br	Br Phase	Bt	Bt Phase	1S Time	Br	Br Phase	Bt	Bt Phase
S1T1	0.7837	113.6	0.116	-60.6	S4T1	0.097	-140.4	0.066	-26.4
S3T3	0.0944	-68.9	0.0358	-136.3	S3T1	0.0005	53	0.0002	73.8
S5T5	0.111	-83.5	0.028	-172.8	S5T1	0.0646	-32.7	0.0361	-139.5
S7T7	0.0119	-100.8	0.008	111.8	S7T1	0.0137	-160.3	0.0127	-16.7
S9T9	0.0348	102	0.012	22.1	S10T1	0.0347	160.7	0.0242	-95.4
S11T11	0.0173	120.7	0.006	15.2	S11T1	0.0454	5.5	0.0356	-82.6
S13T13	0.0105	-105.4	0.004	-179	S13T1	0.035	-166.4	0.031	-67.6
S17T17	0.0065	20.7	0.0027	-66.2	S17T1	0.0241	-26	0.0223	-117.6
S19T19	0.049	95.6	0.003	7.9	S19T1	0.0187	-109	0.019	-17

harmonic component for pole coverage 0.6 has negative phase angle and contributes negative average torque which superimposes on the fundamental component. Thus, fundamental torque is reduced (refer Figure 4.28(c)).

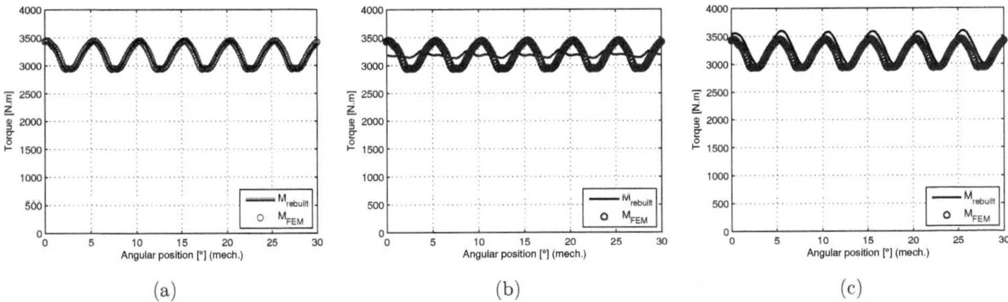

(a)

(b)

(c)

Figure 4.28.: Polar plot D1 pole coverage 0.6, SO $5mm$, $I = I_N$. Torque ripple (a)Basic components. (b)Reduced components (c)Without 3rd.

Load analysis for pole coverage at 0.9:

At pole coverage 0.9 for D2, the relative torque ripple is less. As it can be seen from the polar plot Figure 4.29, the phase angle of the Diagonal elements has been changed and resembles same as D1. The main harmonics that are responsible to generate torque are given in

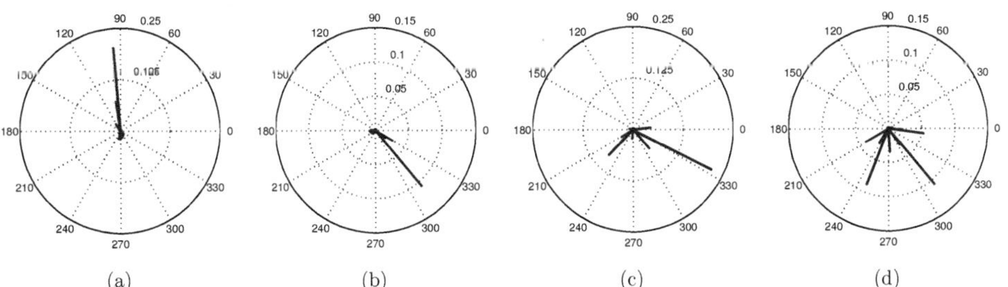

(a)

(b)

(c)

(d)

Figure 4.29.: Polar plot D2 pole coverage 0.9, SO $5mm$, $I = I_N$. (a) Diagonal Elements Radial. (b) Diagonal Elements Tangential (c) First Time Elements Radial. (d) First Time Elements Tangential.

Table 4.13. The magnitude of 1st and 3rd harmonic components are increased. The magnitude of harmonic orders 5th and 7th are reduced. Even harmonics also generate torque pulsation of

Table 4.13.: Diagonal and Time Harmonics D2 pole coverage 0.9, SO 5mm.

Diagonal	Br	Br Phase	Bt	Bt Phase	1S Time	Br	Br Phase	Bt	Bt Phase
S1T1	0.9132	116.5	0.108	-49.7	S4T1	0.087	-134.3	0.053	-7.7
S3T3	0.202	94.9	0.03	-29.9	S3T1	0.0012	91.5	0.0002	-26.1
S5T5	0.0717	99.3	0.018	-36.3	S5T1	0.0605	-48.7	0.0384	-149.8
S7T7	0.0204	122.9	0.007	-42.8	S7T1	0.009	-167.4	0.006	15.2
S11T11	0.0187	-103.9	0.0072	176	S11T1	0.0434	5.6	0.0341	-85.6
S13T13	0.0109	-90	0.009	-177.8	S13T1	0.0338	-148.5	0.029	-53.2
S17T17	0.0121	-57.6	0.0065	149.4	S17T1	0.0263	-29.7	0.0248	-122
S19T19	0.0094	-35.8	0.0053	-127.5	S19T1	0.022	-95.9	0.0212	-4.3

little amplitude due to their low flux density value. As it can be seen from Figure 4.30(a), there is good agreement between the curve from FEM and the reconstructed curve. For this pole coverage, torque harmonics is less and it can be further reduced without harmonics such as 3rd, 4th, 5th and 7th as shown in Figure 4.30(b). Contrary to pole coverage 0.6, average torque due to harmonic pairs $r - S3T3$ and $t - S3T1$ adds to the fundamental torque (refer Figure 4.30(c)).

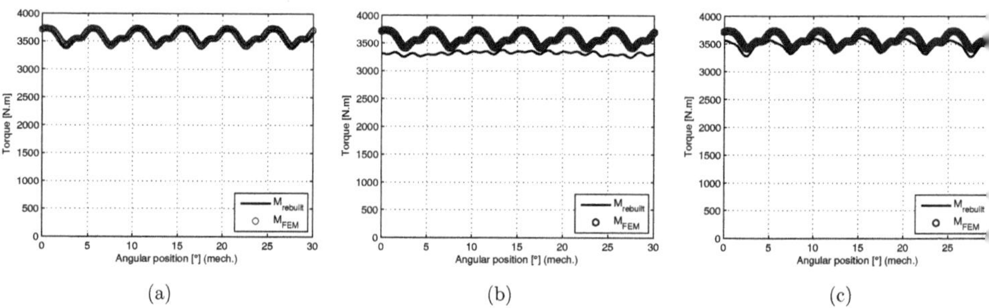

(a)   (b)   (c)

Figure 4.30.: Torque ripple D1 pole coverage 0.9, SO 5mm, $I = I_N$. (a)Basic components. (b)Reduced components (c)Without 3rd.

4.5. Torque Analysis for D3 

D3 arrangement has reduced cogging torque and torque ripple among D1, D2 and D3. Torque analysis for this configuration is made for a slot opening $5mm$ and pole coverage of 0.7.

Under no-load, D3 has the same phasor plot (Figure 4.31) as D1. The phase angle of the diagonal elements that rotate both in space and time are either $+90°$ or $-90°$. Table 4.14

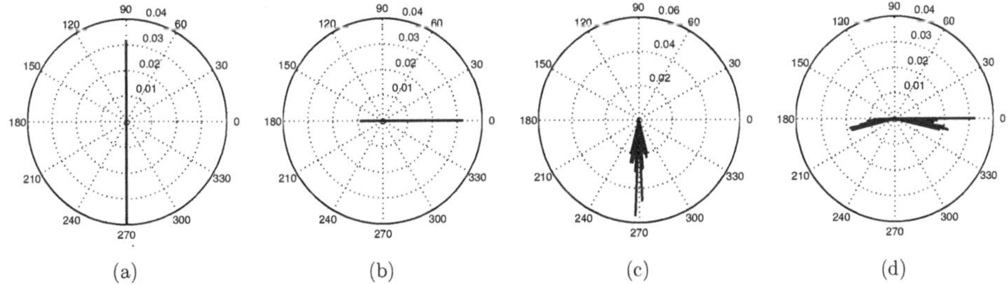

(a)   (b)   (c)   (d)

Figure 4.31.: Polar plot D3 pole coverage 0.7, SO $5mm$, $I = 0A$. (a) Diagonal Elements Radial. (b) Diagonal Elements Tangential (c) First Time Elements Radial. (d) First Time Elements Tangential.

lists the diagonal and time components for magnitude and phase angle at no-load. At 0.7 pole coverage, magnitude of Diagonal harmonics 5th, 7th, 11th and 13th is less. Due to this reason, D3 has reduced cogging torque.

Table 4.14.: Diagonal and Time Harmonics D3 pole coverage 0.7, SO $5mm$.

Diagonal	Br	Br Phase	Bt	Bt Phase	1S Time	Br	Br Phase	Bt	Bt Phase
S1T1	0.8142	90	0.0324	0	S2T1	0.047	-87.7	0.0086	-177.7
S3T3	0.0318	90	0.0022	180	S4T1	0.0556	-92.25	0.017	-2.244
S5T5	0.0294	-90	0.0041	179.9	S5T1	0.0282	-85.5	0.0112	-175.5
S7T7	0.0396	-90	0.0088	179.9	S7T1	0.0299	-94.5	0.0155	-4.5
S11T11	0.0037	-90	0.0013	180	S11T1	0.0191	-81	0.0142	-171
S13T13	0.0058	90	0.0023	-0.04	S13T1	0.025	-99	0.0199	-9
S17T17	0.0022	90	0.0011	0	S17T1	0.0199	-76.4	0.018	-166.5
S19T19	0.0068	-90	0.0038	180	S19T1	0.0199	-103.5	0.0183	-13.5

Under load the appearance of 3rd harmonics is conspicuous. Figure 4.32 gives idea about the polar plot for D3. Compared to D2, the phase angle is almost the same except for the case of 3rd order. Table 4.15 shows the values of Diagonal elements and Time radial elements of D3

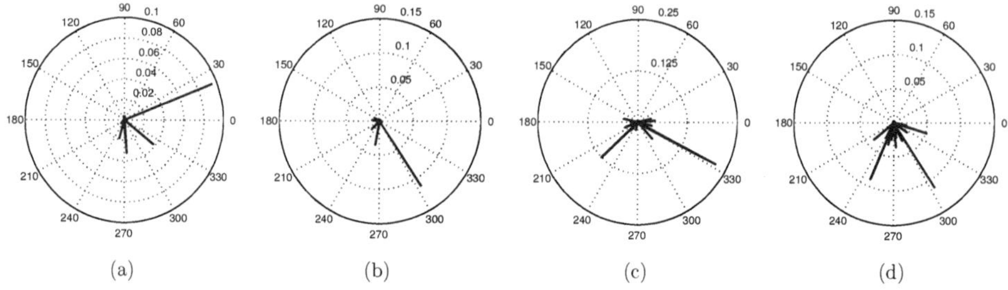

(a)   (b)   (c)   (d)

Figure 4.32.: Polar plot D3 pole coverage 0.7, SO 5*mm*, $I = I_N$. (a) Diagonal Elements Radial. (b) Diagonal Elements Tangential (c) First Time Elements Radial. (d) First Time Elements Tangential.

for pole coverage 0.7. It is evident that D3 has reduced magnitude of harmonics with respect to D1 and D2. The phase angle for the Diagonal elements of 5th and 7th are almost same as D1 an D2. When looking at the torque pulsation for D3, it has less torque ripple which is not

Table 4.15.: Diagonal and Time Harmonics D3 pole coverage 0.7, SO 5*mm*.

Diagonal	Br	Br Phase	Bt	Bt Phase	1S Time	Br	Br Phase	Bt	Bt Phase
S1T1	0.8325	113.7	0.1133	-56.7	S2T1	0.22	-27.4	0.0891	-113
S3T3	0.0945	22.1	0.0345	-100					
S4T7	0.0095	142.3	0.0034	71.5	S4T1	0.1242	-136.5	0.0511	-17.1
S5T5	0.0374	-38.7	0.0162	-134.2	S5T1	0.0538	-47.7	0.038	-140.4
S7T7	0.032	-85.1	0.0108	172.8	S7T1	0.017	-177.5	0.005	-14.3
S11T11	0.0043	-137.8	0.0028	115.8	S11T1	0.0438	4.4	0.0359	-84.9
S13T13	0.0187	112.3	0.0076	19.7	S13T1	0.0337	-158.8	0.0301	-61.3
S17T17	0.0014	-77.8	0.0008	-25	S17T1	0.0247	-26.6	0.0236	-117.8
S19T19	0.0002	-141.7	0.0001	-157.2	S19T1	0.02	-103.1	0.02	-10

the case for D1 and D2. The harmonics that generate torque ripple is evident from Table 4.15. Torque is reconstructed with these harmonics and it follows exactly the same as determined from FEM (refer Figure 4.33(a)). However, if the 3rd, 4th and 5th are neglected for the sake of argument, the torque ripple shown in Figure 4.33(b) can be eliminated.

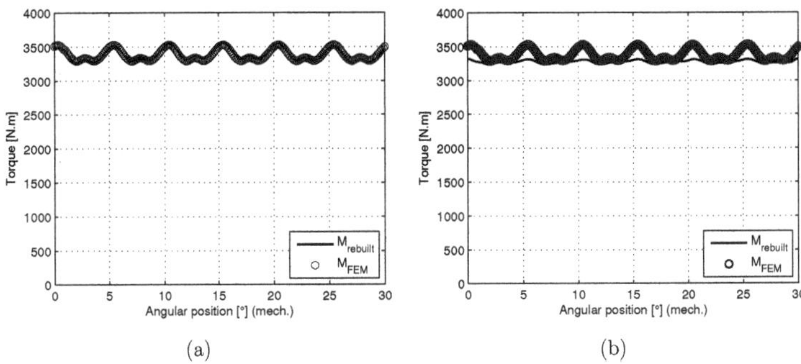

Figure 4.33.: Torque ripple D3 pole coverage 0.7, SO $5mm$, $I = I_N$. Torque ripple (a)Basic components. (b)Reduced components.

4.6. Intermediate Summary

The following points are presented as an intermediate summary for the studied geometries D1, D2 and D3.

- Two different calculations have been studied for geometries D1, D2 and D3, one at no-load and the other at rated load. The harmonics that lie in the diagonal matrix rotate along with rotor both in space and time. For geometries D1 to D3, they have same space and time, for e.g. S5T5.

- The harmonics that are in first time of matrix usually are fixed to stator. These harmonics includes the effect of slots.

- Dominant triplen harmonics, which is third in this case is due to the effect of saturation.

- For geometries D1/D2/D3, the dominant harmonics for parasitic torque under no-load and load are $S5T1$ and $S7T1$.

- During load operation, for geometry D2 and D3, 3rd harmonic component is responsible for small amount of constant torque. This adds to the fundamental constant torque and it is due to the local saturation of iron ring and pole-cap ring. If there is further increase in saturation, the small amount of constant torque will have less influence in the additional torque because the fundamental harmonic which is responsible for fundamental constant torque will be reduced thus reducing the total torque.

- Under load, the flux density which is extracted at the air gap for the torque analysis includes both the effects from stator and rotor fields.

- Thus for geometries D1 to D3 harmonics with lower order are responsible for cogging torque and torque ripple.

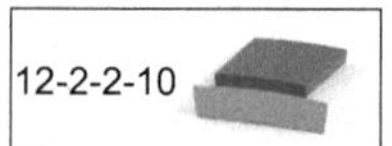

4.7. Torque Analysis for D4

No-load analysis for pole coverage at 0.75:

In this section, the pulsating torque harmonics is derived for D4. As explained in section 4.1.3 of this chapter, torque harmonics order is 12th and 18th for no-load (refer Figure 4.3). However, the 18th order is less. Different harmonics (radial and tangential) are conspicuous at the air

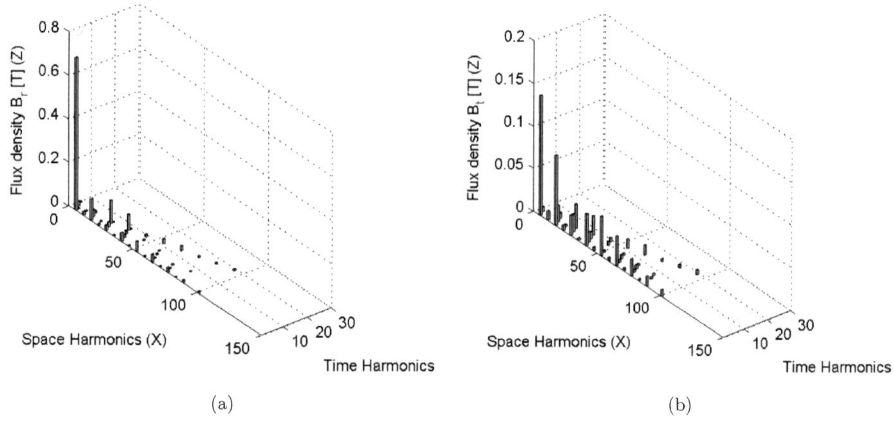

(a)   (b)

Figure 4.34.: Cogging Torque.(a)Radial Component. (b) Tangential Component

gap for pole coverage 0.75 of D4 is shown as bar plot in Figure 4.34. Table 4.16 exhibits some harmonics for D4 arrangement that are responsible for torque pulsation.

In case of D1, D2 and D3, the lower order harmonics, for example the 5th harmonic pair is responsible for the parasitic torque. Interestingly for D4, the harmonics that satisfy the relation 3.80 is not the lower order but the 65th order as mentioned in Table 4.16. It should be noted that if the higher order harmonics are referred to the fundamental order which is 5th

Table 4.16.: Diagonal and Time Harmonics D3 pole coverage 0.75, SO 5*mm*.

Diagonal	Br	Br Phase	Bt	Bt Phase	1S Time	Br	Br Phase	Bt	Bt Phase
S5T1	0.7970	90	0.0426	0					
S25T5	0.0724	-90	0.0192	180	S25T1	0.0005	-94.5	0.0001	179.3
S6513	0.0174	90	0.0088	0	S65T1	0.0276	-101	0.028	-11.2

harmonics for D4 geometry, then it is clear that the torque pulsation producing harmonics pair are comparable to previously studied geometries D1, D2 and D3. The curves in Figure 4.35 shows that only torque reconstructed from pairs $r - S65T13$ and $t - S65T1$ component has good agreement with the results obtained from FEM.

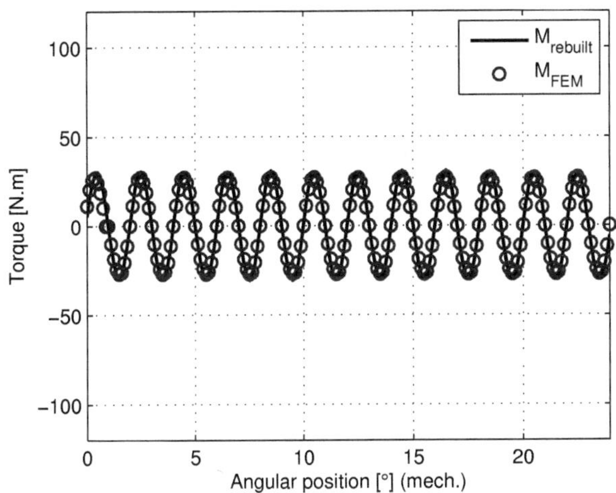

Figure 4.35.: Cogging Torque.

No-load analysis for pole coverage at 0.6:

For the same D4 geometry arrangement, values are listed in Table 4.17 for pole coverage 0.6. This pole coverage has highest relative cogging torque. The 65th harmonic order has changed its phase in this pole coverage. For the purpose of torque reconstruction, the pairs $r - S55T11$, $t - S55T1$ and $r - S65T13$, $t - S65T1$ influencing the pulsating torque are eliminated. Figure 4.36 (b) shows the comparison of pulsating torque. It can be inferred that the higher order components $r - S55T11$, $t - S55T1$ and $r - S65T13$, $t - S65T1$ are also influencing the pul-

Table 4.17.: Diagonal and Time Harmonics D4 pole coverage 0.75, SO 5mm.

Diagonal	Br	Br Phase	Bt	Bt Phase	1S Time	Br	Br Phase	Bt	Bt Phase
S5T1	0.7378	90	0.0404	0					
S55T11	0.0202	-90	0.0083	0	S55T1	0.0212	-78.8	0.0212	-168.7
S65T13	0.0129	-90	0.0068	180	S65T1	0.0264	-101	0.0266	-11.2

sating torque magnitude. Figure 4.36 (a) shows the reconstructed torque from the harmonics mentioned in Table 4.17.

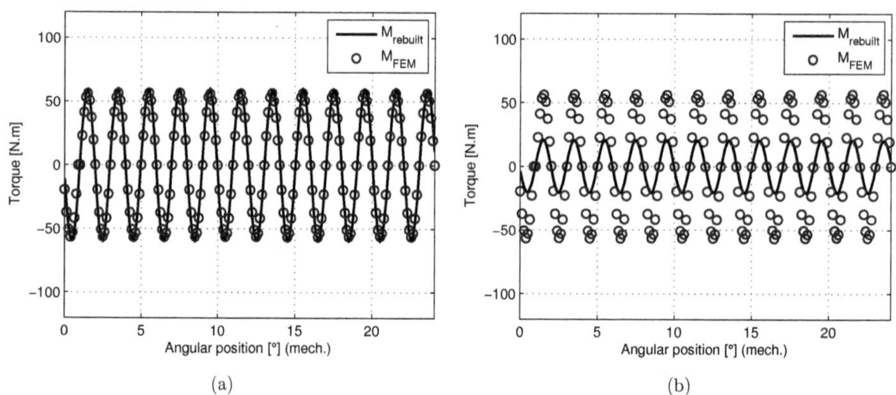

(a)   (b)

Figure 4.36.: Cogging torque D4 pole coverage 0.6, SO 5mm, $I = 0$. Torque ripple (a)Basic components. (b)Reduced components.

Load analysis for pole coverage at 0.7:

Next, the influence of load at rated operation is discussed for D4. The values of Diagonal radial elements and Time space elements for pole coverage 0.7 are tabulated in Table 4.18. Contrary to the no-load, the torque harmonics under load are the 6th and 12th. This is due to the effect of both stator and rotor fields at the air gap, which satisfies the relation in Equation 3.80. In this configuration, the DC shift or constant average torque is from the fundamental radial and tangential component, which is 5th in this case.

The combination to generate torque pulsation are $r - S25T5$, $t - S25T1$ and $r - S35T7$, $t - S35T1$ for 6th torque harmonics and $r - S55T11$, $t - S55T1$ and $r - S65T13$, $t - S65T1$ for 12th torque harmonics. The reconstructed torque due to these components compared with the torque obtained from FEM is shown in Figure 4.37.

Table 4.18.: Diagonal and Time Harmonics D4 pole coverage 0.7, SO $5mm$.

Diagonal	Br	Br Phase	Bt	Bt Phase	1S Time	Br	Br Phase	Bt	Bt Phase
S5T1	0.6977	112.8	0.01405	-47.5					
S25T5	0.1049	-93.6	0.0028	-164	S25T1	0.0003	116.4	0.0051	28.5
S35T7	0.0724	-89.3	0.0021	177.3	S35T1	0.003	-145	0.0026	-64.4
S55T11	0.0223	89.4	0.0106	3.6	S55T1	0.0001	3.9	0.0335	-85.7
S65T13	0.0209	90.4	0.0111	0	S65T1	0.0341	-129.7	0.0345	-39.5

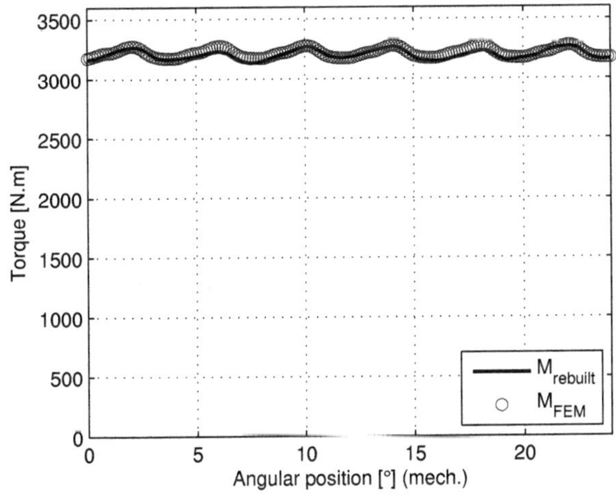

Figure 4.37.: Cogging Torque.

Load analysis for pole coverage at 0.6:

Table 4.19 shows the harmonic components that produce ripple torque of pole coverage 0.6. The fundamental component is less than pole coverage 0.7. The 25th harmonic magnitude is increased and the amplitude of 35th, 55th, 65th harmonics are reduced. Figure 4.38 (a) shows that the identified harmonics fits exactly the results from FEM. The identified harmonics order 25th space and 5th time with 25th space and 1st time, 35th space and 7th time with 35th space and 1st time are eliminated and torque is reconstructed. Figure 4.38 (b) shows that the amplitude of the pulsating torque with the reduced harmonic components. It is evident that the torque ripple is minimized.

Table 4.19.: Diagonal and Time Harmonics D4 pole coverage 0.6, SO $5mm$.

Diagonal	Br	Br Phase	Bt	Bt Phase	1S Time	Br	Br Phase	Bt	Bt Phase
S5T1	0.6624	112	0.01414	-49.2					
S25T5	0.1322	-91.3	0.0329	-173	S25T1	0.0008	-95.4	0.0002	-175.5
S35T7	0.0131	-77.7	0.0039	111	S35T1	0.0026	-157.3	0.0026	-68.7
S55T11	0.0195	92.3	0.0091	-6.6	S55T1	0.0334	6.9	0.0335	-82.7
S65T13	0.0128	-90.5	0.0068	-176.7	S65T1	0.0343	-131.3	0.0345	-41.2

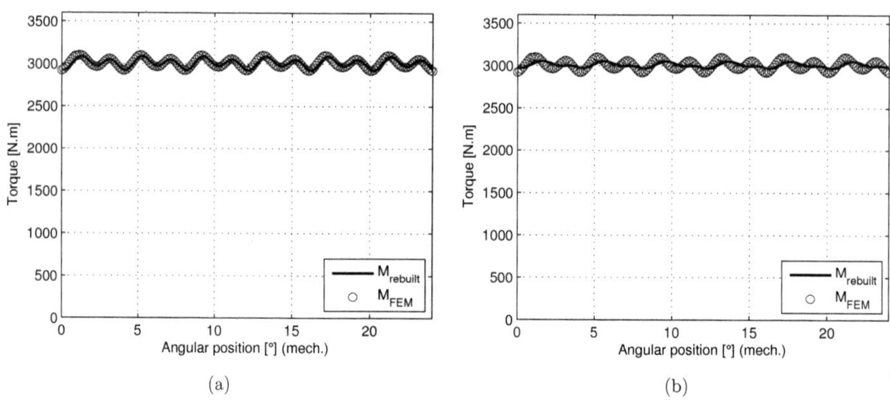

(a)   (b)

Figure 4.38.: Torque ripple D4 pole coverage 0.6, SO $5mm$, $I = I_N$. Torque ripple (a)Basic components. (b)Reduced components.

4.8. Torque Analysis for D5 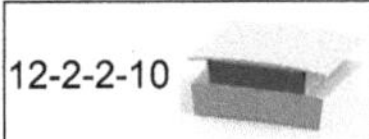

No-load analysis for pole coverage at 0.65:

Cogging torque analysis for D5 is investigated for pole coverage of 0.65. For D5, this pole coverage has high relative torque. The torque harmonic is 12th order similar to D4. Obviously, the harmonics pairs are $r - S55T11$, $t - S55T1$ and $r - S65T13$, $t - S65T1$. This geometry has iron ring which produces additional 25th and 35th harmonics due to local saturation are listed in Table 4.20.

Table 4.20.: Diagonal and Time Harmonics D5 pole coverage 0.65, SO $5mm$.

Diagonal	Br	Br Phase	Bt	Bt Phase	1S Time	Br	Br Phase	Bt	Bt Phase
S5T1	0.7168	90	0.0355	0					
S25T5	0.1335	-90	0.028	180	S25T1	0.0009	-94.3	0.0002	177.6
S35T7	0.0305	-90	0.0068	180	S35T1	0.0002	-95.6	0.0001	174.3
S55T11	0.031	90	0.013	0	S55T1	0.0214	-78.8	0.0165	-168.8
S65T13	0.0036	-90	0.0019	0	S65T1	0.024	-101.2	0.019	-11.2

The influence of higher order harmonics $r - S55T11$, $t - S55T1$ and $r - S65T13$, $t - S65T1$ on pulsating component can be understood from Figure 4.39. It is evident that cogging torque can be reduced, if 55th and 65th harmonic component are not present. Figure 4.39 shows the reconstructed torque when the harmonics listed in the Table 4.20 are considered.

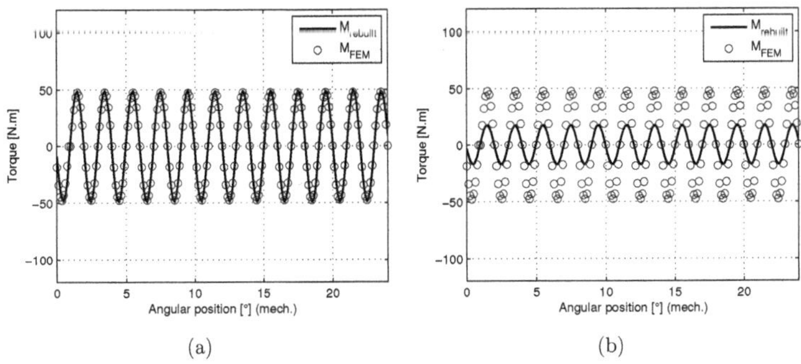

Figure 4.39.: Cogging torque D5 pole coverage 0.65, SO $5mm$, $I = 0$. Cogging Torque (a)Basic components. (b)Reduced components.

<u>No-load analysis for pole coverage at 0.75:</u>

At pole coverage 0.75, the relative cogging torque is very less. It is interesting to note that relative cogging torque is less although the magnitude of 65th harmonic component is higher. This is due to the change in phase angle. Table 4.21 lists the values of D5 for pole coverage 0.75. The comparison of torque obtained from FEM and reconstructed torque considering the harmonics from Table 4.21 is shown in Figure 4.40.

Table 4.21.: Diagonal and Time Harmonics D5 pole coverage 0.75, SO $5mm$.

Diagonal	Br	Br Phase	Bt	Bt Phase	1S Time	Br	Br Phase	Bt	Bt Phase
S5T1	0.7832	90	0.0392	0					
S25T5	0.0957	-90	0.0221	180	S25T1	0.0006	-94.7	0.0001	178
S35T7	0.0816	-90	0.0225	180	S35T1	0.0004	-95.6	0.0001	174.3
S55T11	0.0177	90	0.0076	0	S55T1	0.0213	-78.7	0.0178	-168.8
S65T13	0.0246	90	0.0121	0	S65T1	0.0261	-101.2	0.0216	-11.2
S75T15	0.0127	90	0.007	0	S75T1	0.00006	76.8	0.00005	-13.1

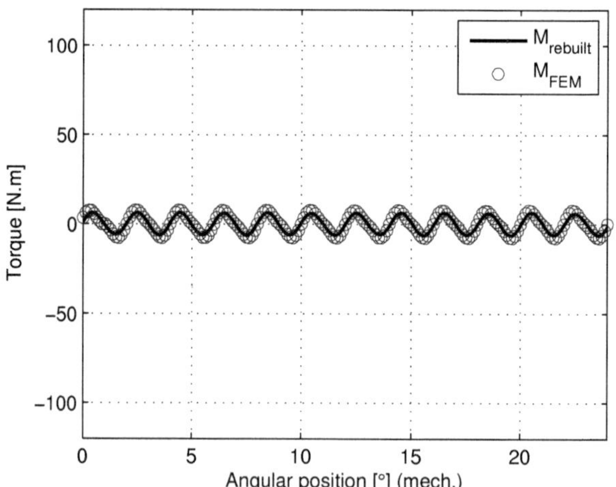

Figure 4.40.: Cogging Torque.

Load analysis for pole coverage at 0.7:

D5 arrangement for pole coverage 0.7 have torque harmonics order 6th and 12th as expected under the rated load current. The values of radial and tangential fields which can produce torque pulsation are listed in Table 4.22. The fundamental component is the 5th. This is increased

Table 4.22.: Diagonal and Time Harmonics D5 pole coverage 0.7, SO $5mm$.

Diagonal	Br	Br Phase	Bt	Bt Phase	1S Time	Br	Br Phase	Bt	Bt Phase
S5T1	0.7442	114	0.1401	-44					
S15T3	0.02	37.3	0.0325	-100	S15T9	0.0002	42.9	0.0002	-89.7
S17T1	0.1216	-125.4	0.0757	-12.6					
S25T5	0.0902	-83.5	0.0259	-151	S25T1	0.0007	-68.3	0.0001	160.7
S35T7	0.0585	-76.8	0.019	-175.7	S35T1	0.0017	-176.6	0.0034	-72.4
S55T11	0.0205	77.4	0.0086	7.5	S55T1	0.0339	1.8	0.0307	-87.9
S65T13	0.0152	108.8	0.0077	14.3	S65T1	0.0328	-125	0.0319	-32

significantly for D5 than D4 at pole coverage 0.7. Harmonics of order 25th, 35th, and 55th have less magnitude than D4.

Figure 4.41 (a) shows that the results obtained from FEM is in close agreement with the results from torque reconstruction. The possible harmonics pairs are $r - S25T5$, $t - S25T1$ and $r - S35T7$, $t - S35T1$ for the 6th torque harmonic and $r - S55T11$, $t - S55T1$ and $r - S65T13$, $t - S65T1$ for the 12th torque harmonic. A constant torque from radial and tangential compo-

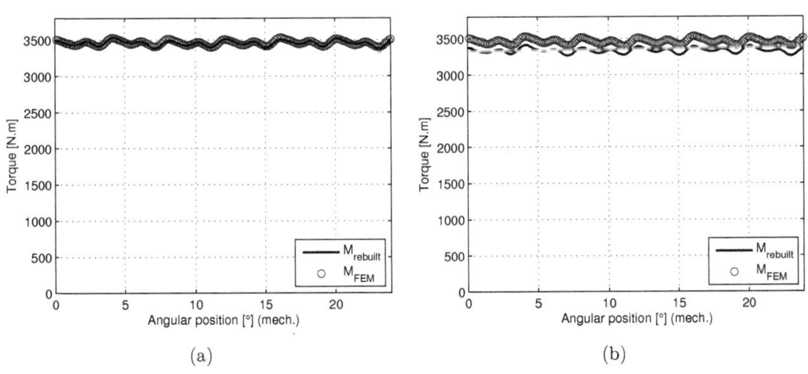

Figure 4.41.: Torque ripple D5 pole coverage 0.7, SO $5mm$, $I = I_N$. Torque ripple (a)Basic components. (b)Reduced components.

nent of field 17th space and 1st time adds with the constant torque from fundamental component. The influence of $r - S17T1$, $t - S17T1$ is depicted in Figure 4.41 (b). The 17th harmonic is present in the stator current sheet distribution. Thus under load, 17th harmonic from the stator interact with 17th harmonic in the rotor field due to local saturation of iron ring, generating constant torque of small amplitude.

<u>Load analysis for pole coverage at 0.55:</u>

Finally, pole coverage 0.55 characteristics is analyzed as it has high relative torque ripple for D5 configuration under load. For this arrangement, 5th and 12th order appears in the torque harmonics. These two order results from harmonics pairs of space order $5^{th}, 7^{th}, 13^{th}, 15^{th}, 17^{th}, 19^{th}, 21^{th}, 25^{th}, 27^{th}, 31^{th}, 33^{th}, 35^{th}, 43^{th}, 45^{th}, 55^{th}$ and 65th with their respective time order. Table 4.23 shows the values of flux density for radial and tangential components.

Table 4.23.: Diagonal and Time Harmonics D5 pole coverage 0.55, SO $5mm$.

Diagonal	Br	Br Phase	Bt	Bt Phase	1S Time	Br	Br Phase	Bt	Bt Phase
S5T1	0.6265	114	0.1394	-49					
S15T3	0.137	-76.4	0.0477	-136.7	S15T9	0.0011	-71.6	0.0003	-136.5
S17T1	0.1237	-129.5	0.0797	-22.8					
S25T5	0.0901	-87.6	0.0217	-177	S25T1	0.0006	-72.4	0.0002	139.2
S35T7	0.0249	85.8	0.01	32	S35T1	0.0012	-158.4	0.003	-85.5
S55T11	0.0108	-130.5	0.0032	162	S55T1	0.0349	9	0.0307	-80.3
S65T13	0.0148	-84	0.0071	-175.3	S65T1	0.0323	-132	0.0313	-40

Fundamental component at pole coverage 0.55 for D5 is less than the pole coverage 0.7. The harmonic order 25th is significantly increased which generates the 6th harmonic torque component. It is interesting to note the influence of 17th order harmonic. Similar to D5 at pole coverage 0.7, 17th order produces constant torque. Figure 4.42 shows the torque pulsation with and without 17th harmonic component. Figure 4.42 (a) represents the torque pulsation with all the harmonic components included.

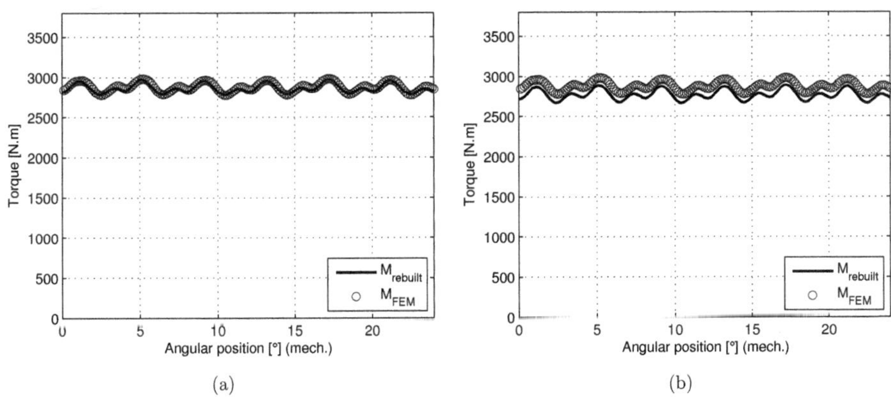

(a) (b)

Figure 4.42.: Torque ripple D5 pole coverage 0.55, SO $5mm$, $I = I_N$. (a) Basic components. (b)without 17th component.

4.9. Torque Analysis for D6

Pulsating torque investigation for D6 has not been performed for all pole coverage variations. It is studied for pole coverage 0.7 and slot opening width $SO = 5mm$. As discussed in section 3.4.3, D6 has less cogging torque and torque ripple of all discussed previous geometries (D1,D2,D3,D4,D5).

Table 4.24.: Diagonal and Time Harmonics D5 pole coverage 0.7, SO $5mm$.

Diagonal	Br	Br Phase	Bt	Bt Phase	1S Time	Br	Br Phase	Bt	Bt Phase
S5T1	0.7503	90	0.0372	0					
S15T3	0.0061	90	0.006	180	S15T9	0.00004	88	0.00004	178
S25T5	0.035	-90	0.0062	180	S25T1	0.0002	-94	0.00003	176.7
S35T7	0.0328	-90	0.0085	180	S35T1	0.0002	-95.6	0.00004	174.4
S55T11	0.0035	90	0.0014	0	S55T1	0.00007	80.5	0.00003	170.5
S65T13	0.0056	90	0.0011	0	S65T1	0.0244	-101.3	0.0214	-11.2

Cogging torque for D6 is approximately $1N.m$ peak to peak torque. Table 4.24 lists the few harmonics of radial and tangential components. The amplitude of the tangential harmonics in time elements are less compared to the other geometries. The results of the torque obtained from FEM and reconstructed torque using harmonics at no-load are shown in Figure 4.43. The curves show that the selected harmonics have good agreement.

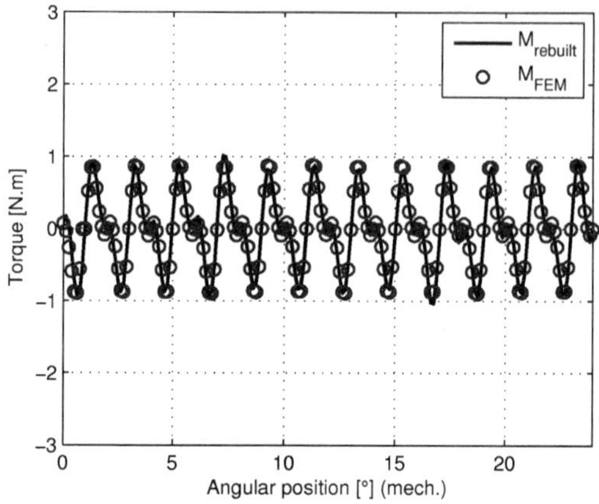

Figure 4.43.: Cogging Torque.

Torque ripple for D6 is studied for pole coverage of 0.7. Due to the stator field, the phase angle is displaced. The values of harmonics magnitude and phase angle are provided in Table 4.25.

Table 4.25.: Diagonal and Time Harmonics D5 pole coverage 0.7, SO 5*mm*.

Diagonal	Br	Br Phase	Bt	Bt Phase	1S Time	Br	Br Phase	Bt	Bt Phase
S5T1	0.7056	113.7	0.1395	-45.5					
S15T3	0.074	11.3	0.0421	-97.5	S15T9	0.0005	11.2	0.0003	-100
S17T1	0.1458	-137.6	0.0723	-9.8					
S25T5	0.033	-75	0.013	-140.4	S25T1	0.0002	-79.2	0.0003	-100.4
S35T7	0.0285	-100.8	0.0099	162	S35T1	0.0023	169	0.0021	-79
S55T11	0.0026	105	0.0014	14	S55T1	0.0333	3.8	0.0309	-85
S65T13	0.0038	77.8	0.0022	-16.3	S65T1	0.0314	-126.4	0.0304	-32.6

As explained for geometry D5, the 17th harmonic component for D6 also contributes to a DC shift in the average torque. This effect is shown in Figure 4.44 (b). Figure 4.44 (a) proves that the reconstructed torque with harmonics from the selected order are comparable with the torque ripple from FEM.

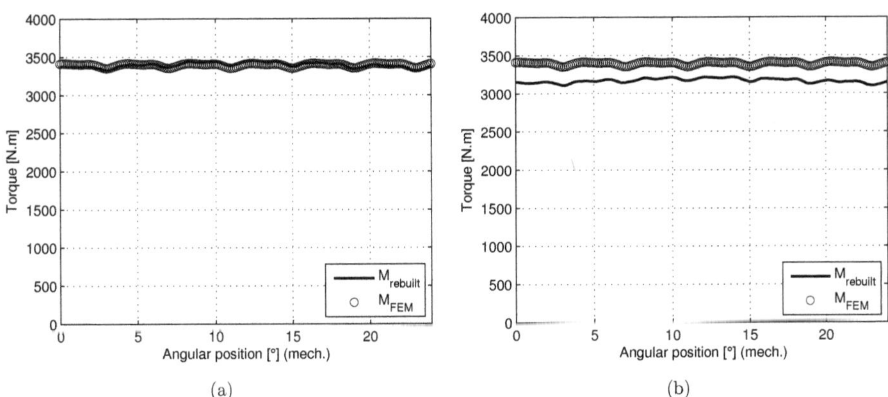

Figure 4.44.: Torque ripple D6 pole coverage 0.7, SO $5mm$, $I = I_N$. Torque ripple (a)Basic components. (b)Reduced components.

4.10. Summary

The 2D FFT analysis used in this chapter helps in understanding about the harmonics that are responsible for the generation of pulsating torque. Based on the analysis performed in this chapter, the following points are summarized.

- Two different calculations have been studied for geometries D1, D2 and D3, one at no-load and the other at rated load. The harmonics that lie in the diagonal matrix rotate along with rotor both in space and time. For geometries D1 to D3, they have same space and time, for e.g. $S5T5$.

- The harmonics that are in first time of matrix usually are fixed to stator. These harmonics includes the effect of slots.

- Dominant triplen harmonics, which is third in this case is due to the effect of saturation.

- For geometries D1/D2/D3, the dominant harmonics for parasitic torque under no-load and load are harmonic pairs from 5th and 7th.

- For geometry D4/D5/D6 under no-load, the dominant harmonic pairs are 55th and $75^{th}5$,

- For geometry D4/D5/D6 under load, the dominant harmonic pairs for parasitic torque are 25th, 35th, 55th and 75th.

- Thus for geometries D1 to D3, harmonics with lower order are responsible for cogging torque and torque ripple. It should be noted that if the higher order harmonics for D4 to D6 geometries are referred to the fundamental order which is the 5th harmonics, then it is clear that the torque pulsation producing harmonic pairs are comparable to the geometries D1, D2 and D3. It clearly shows that only lower order harmonics generate pulsating torque.

- Under load, the flux density which is extracted at the air gap for the torque analysis includes both the effects from stator and rotor fields.

- During load operation, for geometries D2 and D3, 3rd harmonic component is responsible for small amount of constant torque. This adds to the fundamental constant torque and it is due to the local saturation of iron ring and pole-cap ring. If there is further increase in saturation, the small amount of constant torque will have less influence in the additional torque because the fundamental harmonic which is responsible for fundamental constant torque will be reduced thus reducing the total torque.

- During load operation, similar to geometries D2 and D3, geometries D5 and D6, 17th harmonic component is responsible for small amount of constant torque. This adds to the fundamental constant torque and it is due to the local saturation of iron ring and pole-cap ring.

5. Measurement Results and Parasitic Torque Special Study

One of the geometries *D3 arrangement*, discussed in the previous chapter has been implemented as prototype. The design for geometry D3 was performed at ***Institut for Elektrische Maschinen, Antriebe und Bahnen,(IMAB) TU-Braunschweig***. The complete test bench was implemented at ***AS Drives and Services GmbH***. At the same time, the inverter for the prototype was developed and tested at IMAB as well as integrated in the test bench as a complete direct drive topology. This chapter concentrates on the test bench set-up and measurements obtained from the test bench. Additionally, this chapter also discusses about parasitic torque for an optimized machine considering operating points such as field weakening and half load $I_N/2$.

5.1. Measurement Set-up

From the results of measurements, a quantitative validation about the characteristics of the designed machine can be performed. The measurements are done using the test bench facility at AS Drives and Services GmbH. The measured data of cogging torque and torque ripple are provided by the AS Drives and Services GmbH. The schematic diagram of the test bench is shown in Figure 5.1. The test bench consists of a direct-drive (Permanent Magnet Synchronous Machine) coupled to an asynchronous machine (ASM 4-pole) through a Flender spur gear system (type:H2 PH 05 B). The direct drive operates in motor mode and ASM operates in the generator mode. The inverter from IMAB is used to control the drive. An additional inverter from SEW delivers the generated power from ASM back to the common DC-Link. The Direct drive consists of a hollow type rotor, which sits directly on the cylinder as described in chapter

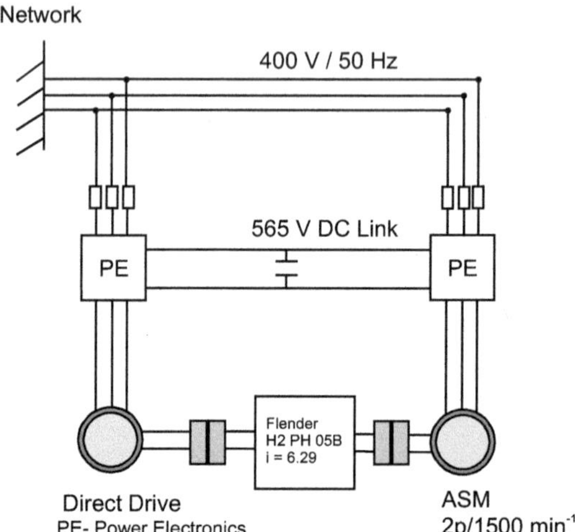

Figure 5.1.: Schematic diagram of Set-up.

2. The machine is water cooled and the design current density is $J = 6 \ A/mm^2$. Information about motor cooling system is out of scope of this thesis work. However, it should be noted that the machine is within the thermal operating limits during rated operation. In the prototype, the drive control parameters can be tested and tuned close to the real operating and environment conditions. Another advantage with the test bench is that the machine and the inverter can be tested for over-load conditions also. The torque is measured using the measurement device T10 with a precision of 0.03 - 0.05 %. The data acquisition such as torque, speed, efficiency, temperature etc. are done with Lab View. Even-though many performance measurements can be done with the test bench, in this work only important measurements such as no-load torque and load torque closely related to this work are considered.

5.1.1. Pulsating Torque

The measurement of cogging torque and torque ripple performed for geometry D3 using Samarium Cobalt magnets. During the design stage, the machine is calculated for NdFeB magnet type. Other material parameters that are used for the design are given in Table 2.1 in Chapter 2.5. For the convenience of data acquisition, the direct-drive test motor is driven by ASM close to zero speed for both no-load and load conditions. The characteristics of parasitic torque should be the same whether the direct runs at $1min^{-1}$ or at $200min^{-1}$.

Figure 5.2.: Test Bench [AS Drives & Services GmbH]

The cogging torque comparison of measured results and the calculated results from FEM is shown in Figure 5.3. The calculated values have the same torque pulsating form as the measured values for one electrical period. During measurement, there is DC component in the torque and this is due to the sealings of the rotor. To have better visualization, the DC shift is subtracted from the measured result.

For one electrical period, the pulsating torque at no-load have *six* periods for both calculated and measured values and they are in very good agreement. The interpretation of the pulsating torque under no-load should be referred to section 4.5 in the previous chapter. The harmonics pairs that generate the cogging torque are $r - S5T5$, $t - S5T1$, $r - S7T7$, $t - S7T1$, $r - S11T11$, $t - S11T1$ and $r - S13T13$, $t - S13T1$.

In case of load operation, the measured torque (blue curve) for stator current 200 A effective, the calculated torque from FEM (black curve) for stator current 200 A effective and the calculated torque from FEM (red curve) for a rated current of 177 A effective are shown in Figure 5.4. The torque characteristics between the measurement and the calculation are similar. The average torque of the measurement is less compared to the FEM at 200 A rated current. The reason is that the permanent magnets used in the prototype is Samarium Cobalt(SmCo) that

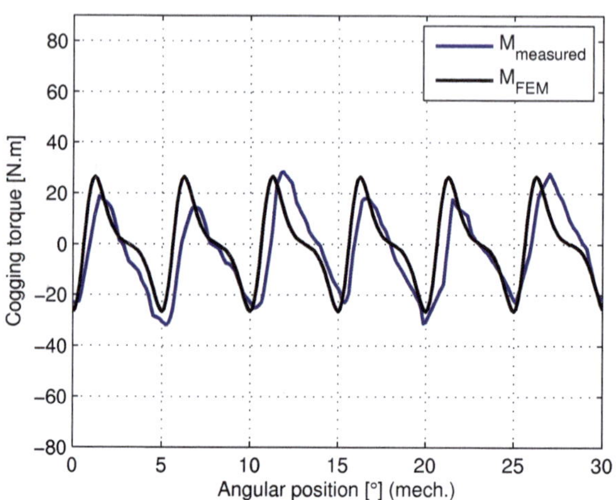

Figure 5.3.: Comparison Measurement and FEM calculation - Cogging Torque.

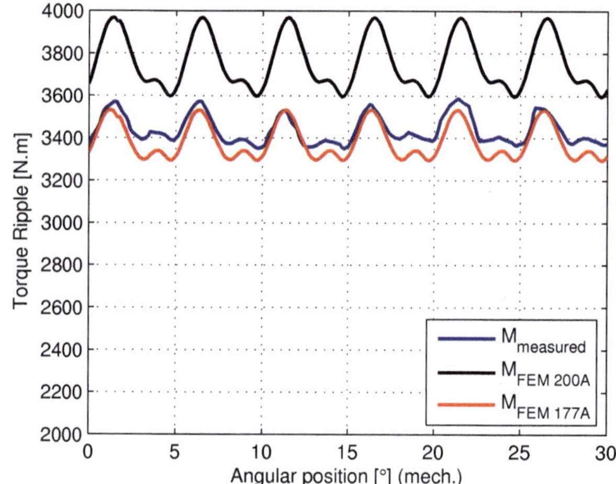

Figure 5.4.: Comparison Measurement and FEM calculation - Torque Ripple.

has different magnetic properties than Neodym Boran (NdFeB) magnets, which is used in the design calculation. At room temperature, SmCo has remanence flux density around 1 T and NdFeB remanence flux density is 1.2 T. The difference is 20% which has impact on torque. The other reason could be the parasitic air gap. Figure 5.5 shows the lamination used for the

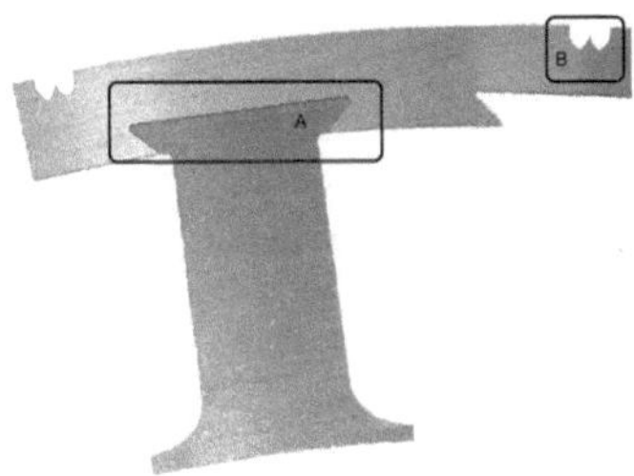

Figure 5.5.: Stator Lamination.

prototype. The teeth is stamped separately and the coils are wound around the teeth and inserted in to the yoke lamination which has wedge form as shown in place A in Figure 5.5. During assembly of laminations (yoke and teeth) there would be formation of parasitic air gap approximately between 0.2 and 0.5 mm which forms magnetic reluctance in the magnetic circuit path. Moreover, at place B in the yoke there is another wedge to stack the lamination together to the stator frame. Thus, the yoke height is reduced at this position which may lead to saturation of magnetic circuit at yoke. These effects will have impact on the torque of the machine. Thus there is considerable difference in the machine when compared to the calculated results. Interesting to see is the pulsating torque behaviour for the considered stator currents. It is evident from the curves in Figure 5.4 that the harmonics which is interpreted for D3 arrangement in the previous chapter is valid for 200 A stator current. The working harmonics which account for pulsating torque are $r - S3T3$, $t - S3T1$, $r - S4T1$, $t - S4T7$, $r - S5T5$, $t - S5T1$, $r - S7T7$, $t - 75T1$, $r - S11T11$, $t - S11T1$ and $r - S13T13$, $t - S13T1$.

5.2. Special Study

In the following sub-sections, further extension of the torque analysis has been performed as special study. This study includes pulsating torque using optimization method, torque ripple in field weakening range and torque ripple under half load operating point. It should be noted that not all the geometries are covered for special study. The studied work in this section are as follows,

- In the optimization method, two different possibilities for D1 and D3 geometry are studied under no-load.

- Field weakening for D1 geometry.

- Half load operation for geometries D1,D3,D4 and D6.

5.2.1. Introduction to Optimization Methods

The details of optimization method in the following section is from author's own publication in institutes's annual work report [67]. Although optimization techniques are widely implemented in various fields, they are also being used by the electrical machine design engineers to get insight about the magnetic circuit. The magnetic circuit parameters are very sensitive and they influence the output performance such as torque ripple, losses, efficiency and cost of the electrical machine. Thus, optimization methods find a solution in the design criterion by fulfilling the given objective function either to minimize or to maximize under constrained or unconstrained geometrical parameters. For example, the objective function can be to minimize torque pulsations or to maximize the efficiency, etc. When performing an optimization for electrical machines, their parameters can be directly optimized with saturation dependent numerical calculations provided there is a possibility to combine FEM and optimization tools. In this work, the FEM software is FLUX, CEDRAT [60] and the optimization tool is FGot from G2ELAB [68].

5.2.2. Optimization Methods and Model Parameters

Basically, the optimization methods are classified into deterministic and stochastic methods. The optimization methods algorithm are understood from the tree shown in Figure 5.6. The theory behind the optimization techniques are found in the literature [5]. The mathematical explanation on minimizing the objective function with number of variable vectors $\overline{X}$ can be written as

$$minimize \quad f(\overline{X})$$
$$g_i(\overline{X}) \leq 0, \quad 1 \leq i \leq m$$
$$h_i(\overline{X}) \leq 0, \quad 1 \leq i \leq k$$

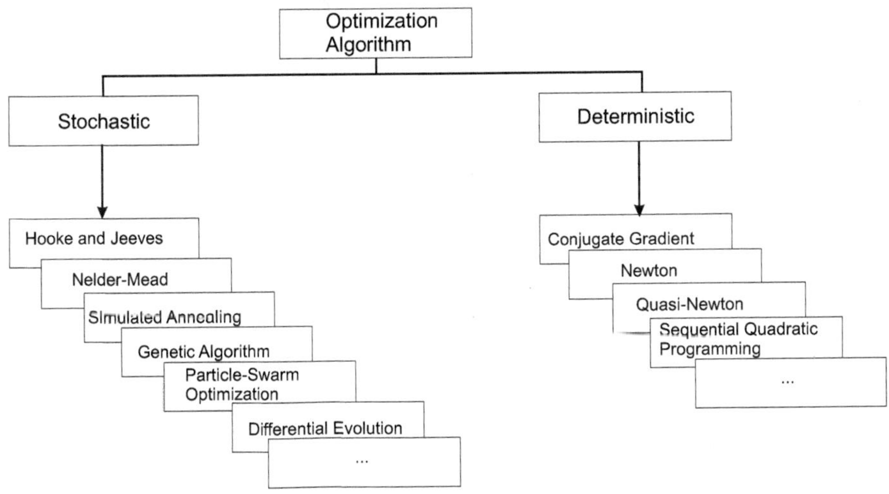

Figure 5.6.: Optimization Methods.

The objective function can be a single or a multi-objective function, with g_i and h_i are equality and inequality constraints. The deterministic methods are local optimum methods that use gradient methods to find the objective function, whereas the stochastic methods find the global optimum based on evolution methods. There are also algorithms combining both methods [18]. All the methods should be carefully interpreted. Sometimes, the convergence is poor in local optimum. This can be improved with the help of good initial values. On the other hand, other methods based on evolutionary strategy are able to find global optimum. In this work, a Genetic Algorithm based on evolution strategy with mono-objective function is illustrated. When performing an optimization for electrical machines, the following steps are necessary:

- Selection of variables or geometry dimensional parameters to be optimized. These are number of poles, height and width of slot, outer radius of stator or inner radius of rotor, flux density in teeth, etc.

- Based on physical limits, constrained or unconstrained limits for the parameters should be set.

- Providing good initial values for the parameters based on experience.

- Preparing the objective function for which the variables should be optimized.

- Selecting the optimization algorithm to perform optimization.

5.2.3. Optimized Geometry

The idea of using the optimization problem is to vary the air gap field and the objective function is to minimize the pulsating torque. The points that are varied during the optimization process are shown in Figure 5.7. These points are varied towards radial direction. The constraints are made to these points such that they are within the geometrical limits. For example, upper limits and lower limits of the parameters should be given. Of course, there are other parameters such as slot opening, pole coverage, teeth width, yoke height etc., which can be handled in the optimization problem. Such problems of finding a solution is beyond the scope of this work.

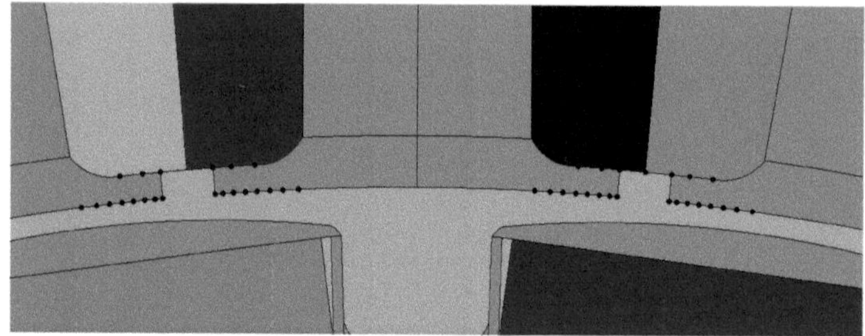

Figure 5.7.: Points for Optimization.

It should be noted that performing optimization with FEM tool is a time consuming process. The FEM calculation carried out in this work is 2D problem for one sixth of an elementary machine (one period of torque pulsation is $5°mech.$). Computational time to vary the points shown in Figure 5.8 to minimize cogging torque with 32 GB RAM work station is approximately 20 hours. Two variants of geometry dimension shown in Figure 5.8 are discussed for torque analysis. For example, from the manufacturing point of view Variant 1 can be made such that the laminations are stamped by welding on the surface of the air gap with small notch. Variant 2 in Figure 5.8 (b) is an intermediate result of optimization algorithm. Similar geometry as Variant 2, stator tooth optimization of an induction machine is presented for minimum torque ripple [69]. The minimum torque pulsation for such tooth modification in this optimization study is obtained in variant 1 and not in variant 2.

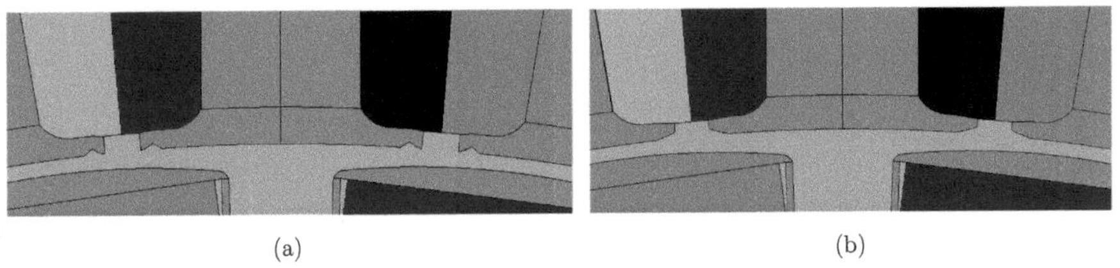

(a)   (b)

Figure 5.8.: Optimized Geometry (a)Variant 1. (b)Variant 2

5.2.4. Torque Analysis Optimized Geometry

In this section, torque analysis for optimized geometry D1 and D3 for two variants under no-load is discussed. Table 5.1 lists the air gap radial and tangential components for the variant 1 of D1 optimized geometry. The working harmonics which are responsible for the pulsating

Table 5.1.: Diagonal and Time Harmonics D1 pole coverage 0.7, SO $5mm$.

Diagonal	Br	Br Phase	Bt	Bt Phase	1S Time	Br	Br Phase	Bt	Bt Phase
S1T1	0.7896	90	0.0333	0	S4T1	0.0298	-92.3	0.0251	-2.3
S5T5	0.1191	-90	0.0232	180	S5T1	0.0182	-85.5	0.017	-175.5
S7T7	0.0854	-90	0.0193	180	S7T1	0.0225	-94.5	0.0213	-4.5
S11T11	0.0299	90	0.0100	0	S11T1	0.013	-81	0.0129	-171
S13T13	0.0282	90	0.0115	0	S13T1	0.0185	-99	0.0188	-9
S17T17	0.01	-90	0.0054	180	S17T1	0.0156	-76.5	0.016	-166.5
S19T19	0.0118	-90	0.0067	180	S19T1	0.0157	-103.5	0.016	-13.4

torque are $r - S5T5$, t $S5T1$, $r - S7T7$, $t - S7T1$, $r - S11T11$, $t - S11T1$ and $r - S13T13$, $t - S13T1$. Figure 5.9 shows the cogging torque comparison for Variant 1 and Variant 2 for one elementary machine. In Variant 1, the cogging torque magnitude of optimized geometry (curves $M_{rebuilt}$, M_{FEM}) is reduced considerably when compared to the original geometry D1 (red curve M_{D1}). Variant 2 is one of the variations during the optimization program process. Variant 2 does not make any changes in the cogging torque. Figure 5.10 shows the use of optimization methods for geometry D3. The harmonics for geometry D3 are not tabulated. It is evident from Figure 5.10 that there is no change in the cogging torque for this geometry because there is no reduction in magnitude of harmonics in D3 structure after optimization. Moreover, the THD factor is less as shown in the Table 3.9 for D3 structure.

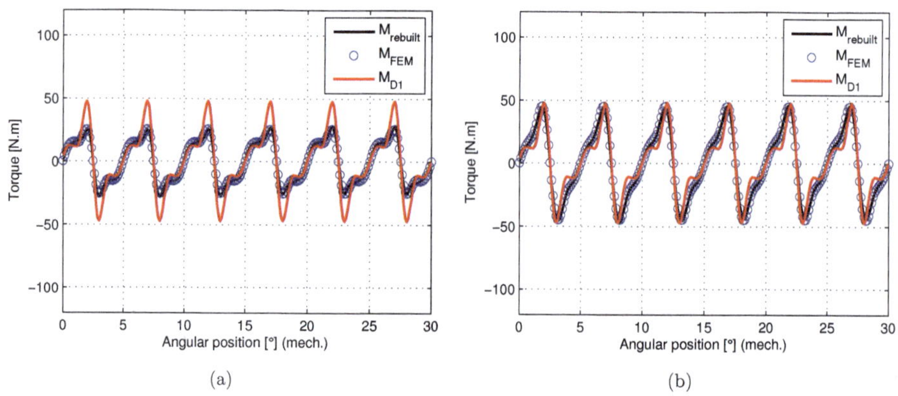

Figure 5.9.: Cogging Torque D1 0.67, SO $5mm$ (a)Variant 1. (b)Variant 2.

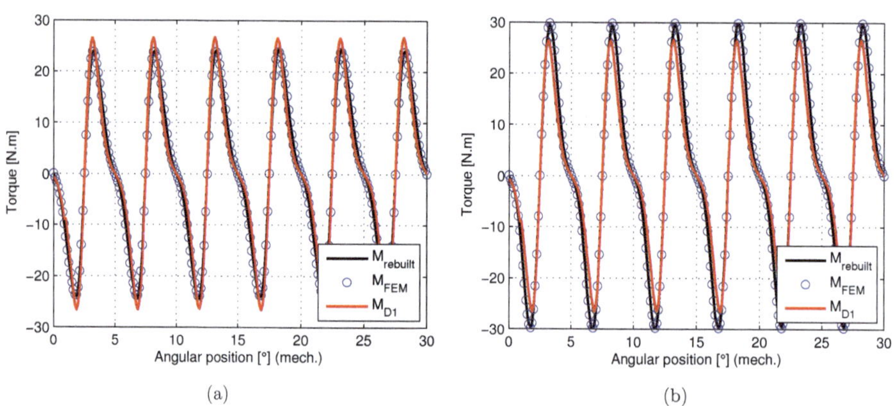

Figure 5.10.: Cogging Torque D3 0.67, SO $5mm$ (a)Variant 1. (b)Variant 2.

5.3. Torque Ripple under Field Weakening

The theory about the field weakening range is explained in Chapter 2.5. For permanent magnet synchronous machine, field weakening is done by imposing negative d-axis current. In the thesis work [70], the influence of torque ripple in the field weakening range is more extensively studied. The machine working in the field weakening range has increased torque ripple. The field weakening angle β which is explained in chapter 2.4.1 is selected according to the MTPA line.

In this work, D1 geometry is analysed for a field weakening at speed of $400min^{-1}$, which is twice the rated speed. At this speed, the power is kept constant and the average mechanical torque according to Equation 2.1 is reduced by half. The radial and tangential components of air gap flux density is tabulated in Table 5.2. The dominant harmonics pairs for the pulsating

Table 5.2.: Diagonal and Time Harmonics D1 pole coverage 0.7, SO $5mm$.

Diagonal	Br	Br Phase	Bt	Bt Phase	1S Time	Br	Br Phase	Bt	Bt Phase
S1T1	0.626	104.5	0.1031	-29					
S5T5	0.1255	-90.5	0.0257	-176	S5T1	0.0308	27.2	0.0284	-62.5
S7T5	0.0766	-88.4	0.0213	173.7	S7T1	0.0457	-102.9	0.0441	-12.9
S11T11	0.0343	89.4	0.0133	2.6	S11T1	0.0231	51.3	0.0232	-38.2
S13T13	0.0249	90.8	0.0108	-2.5	S13T1	0.0406	-130.6	0.0409	-40 4
S17T17	0.0133	-90.3	0.0071	-178	S17T1	0.0132	-18.8	0.0134	-109.2
S19T19	0.0103	-89.4	0.0059	178	S19T1	0.0241	-104.6	0.0244	-14.5

torque in field weakening speed of $400min^{-1}$ are $r - S5T5$, $t - S5T1$ and $r - S7T7$, $t - S7T1$. However, the pairs $r - S11T11$, $t - S11T1$ and $r - S13T13$, $t - S13T1$ are influencing the phase change of the curve. Figure 5.11 (a) shows the reconstructed torque compared with the torque obtained from FEM with working harmonics pairs such as $r - S5T5$, $t - S5T1$, $r - S7T7$, $t - S7T1$, $r - S11T11$, $t - S11T1$ and $r - S13T13$, $t - S13T1$. Figure 5.11 (b) shows the influence of $r - S11T11$, $t - S11T1$ and $r - S13T13$, $t - S13T1$.

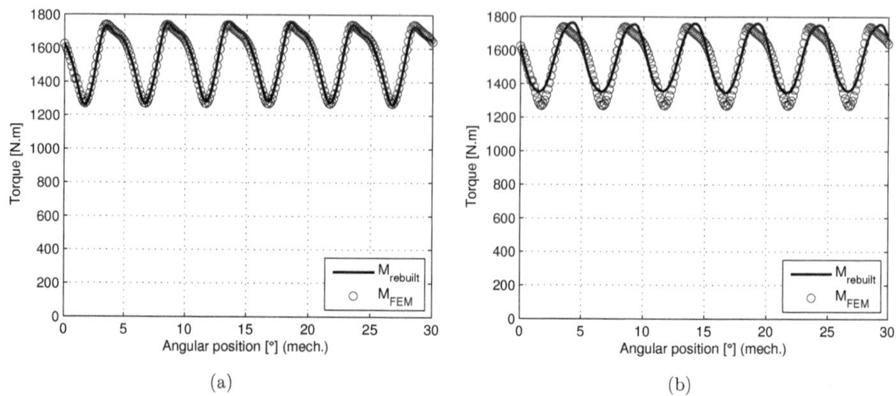

Figure 5.11.: Torque Ripple field weakening (a) with 11th. (b)without 11th

5.4. Torque Ripple under Alternative Field Weakening

In the classical approach of field weakening as detailed in the previous section, there are copper losses in the stator due to the presence of negative d-current. Moreover, the converter has to be over dimensioned to bring the reactive power in to the machine. There are other alternative approaches for PMSM to operate in the field weakening range. These are discussed in many publications [71], [72]. In connection to this work on torque ripple, field weakening by mechanical axial shift of the radial flux PMSM machine is studied. Geometry D1 is considered as example.

Figure 5.12 shows the CAD geometry of axial shift for geometry D1 where half the rotor is moved axially. This mechanical shift is explained in the paper [71]. Other idea is to displace the rotor on both sides which is not shown in the figure. The rotor is made into two pieces axially and connected to the shaft. The splitting of the shaft into two parts mechanically can be realized by having a construction with fluid in the middle. This fluid will rotate along with rotor and after certain speed centrifugal force is exerting the fluid to move the rotor axially. The design of such mechanism is beyond the scope of this topic and it is not further explained. Figure 5.13 shows the CAD model of the geometry D1 with stator coils in the slots. It is an elementary machine and there are three coils belonging to each phase. It can be seen in Figure 5.13 that the coils go around the tooth. Hence, this winding has the name *Tooth-Coil Winding*.

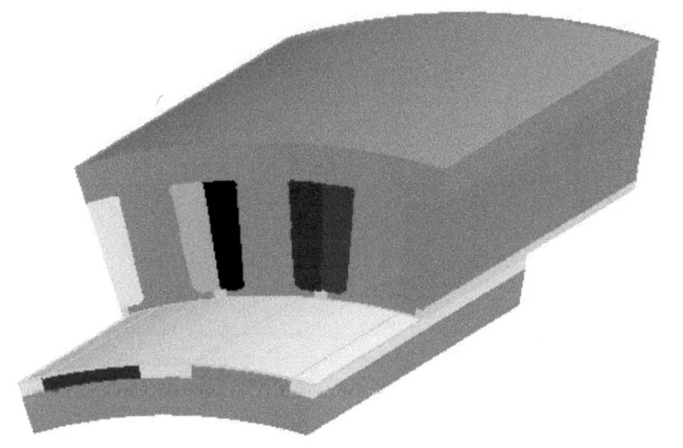

Figure 5.12.: Axial Shift Realization.

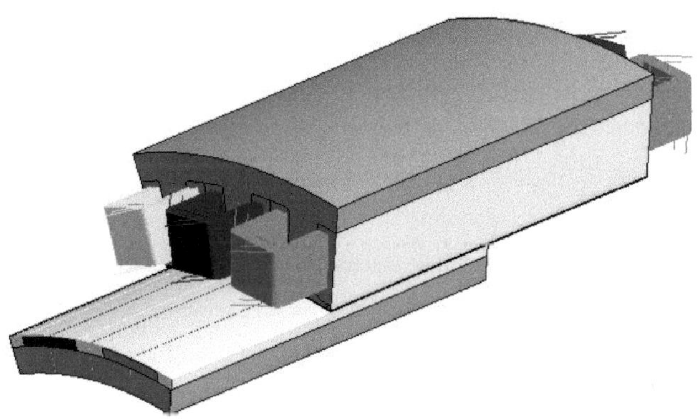

Figure 5.13.: Geometry with Stator Coils.

The axial force that needs to displace the rotor magnetic circuit from stator magnetic circuit is obtained from three dimensional FEM calculation. The calculated force and axial displacement in length (mm) of the rotor is shown in Figure 5.14. The required force to displace half the length of the rotor $112.5mm$ is approximately $1kN$. Coming to the pulsating torque calculation, Figure 5.15 shows the cogging torque and the torque ripple after the rotor displacement. Due to huge computation time in 3D FEM calculation, the rotor rotation is made for 5° mechanical. The pulsating torque periodicity 5° mechanical and this periodicity is sufficient to evaluate the pulsating torque. Figure 5.15(a) shows the cogging torque for one torque periodicity. There

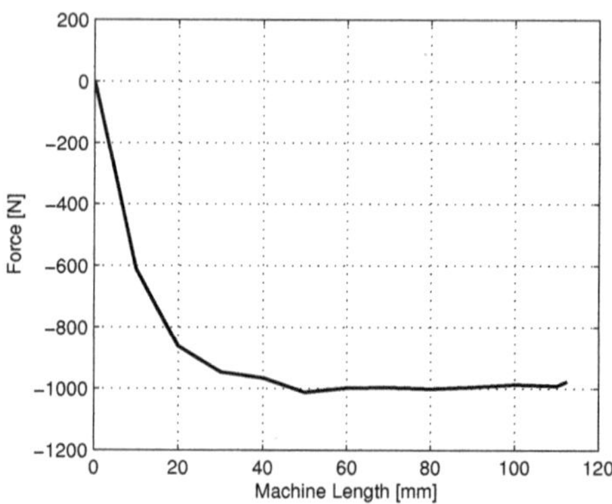

Figure 5.14.: Axial Force vs Displacement.

are small ripples in the curve and this is due to the numerical error from the FEM program because the mesh used for 3D calculation is an automatic mesh from the program which is not evenly distributed in the air gap. Cogging torque requires mesh in the air gap as explained in the paper [61]. Figure 5.15(b) shows the torque ripple with half the displacement of rotor. Half the displacement of the rotor means half the air gap flux density linking the stator. The synchronous generated voltage according to the last term of Equation 2.9 is reduced to half considering it as proportional to flux. The machine now runs within the stator voltage limit and the torque is halved. Figure 5.16 shows the comparison of relative values of the cogging torque and torque ripple. In both the cases, the torque pulsation magnitude is reduced approximately half the value in the case of axial shift of the rotor compared to the field weakening method using negative d-axis current. If the mechanical shift of the rotor really benefits by using it practically, then it can be applied. However, other criteria for example such as bearings dimensions, axial forces from the magnetic system, maintenance, natural frequency of the rotor, etc. have to be taken care during construction of the machine.

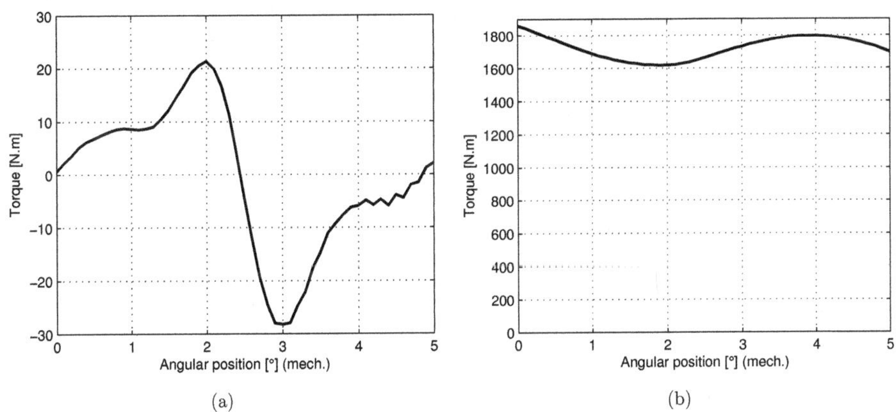

Figure 5.15.: Pulsating torque field weakening in axial shift (a) Cogging torque. (b)Torque ripple.

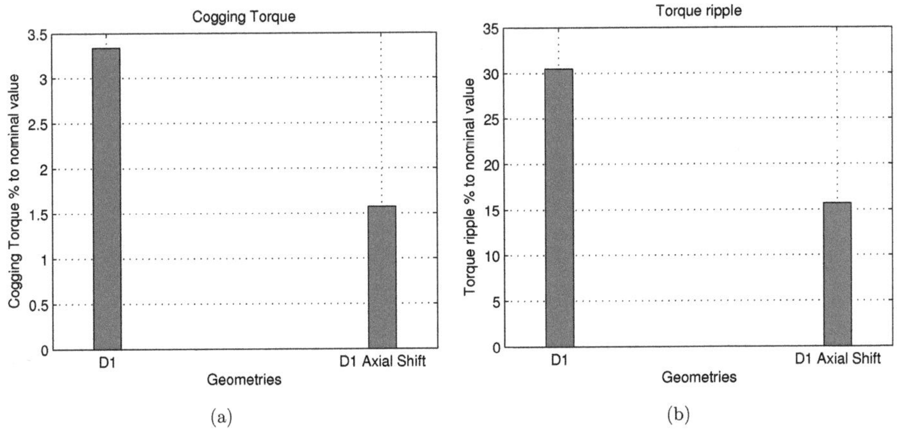

Figure 5.16.: Pulsating torque relative value comparison (a) cogging torque. (b)torque ripple.

5.5. Torque Ripple under Half Load Nominal Operating Point

The final part of this special study is to analyze torque ripple of the machine under half load condition. Under half load, the machine is not saturated. Thus the half load analysis is performed to study the impact of harmonics due to saturation in the machine. The half load torque ripple study is made for geometries D1, D3, D4 and D6. The D1 and D4 geometries are investigated for the pole coverage of 0.6 where maximum torque ripple occurs and the other pole coverage of 0.7 where torque ripple is less than the pole coverage 0.6. For the geometries D3 and D6, due to geometry limitation the investigation is made for pole coverage of 0.7.

5.5.1. Torque Analysis at Half Load D1

Half load analysis for pole coverage at 0.6:

The working harmonics that generate torque ripple for D1 at pole coverage 0.6 are listed in Table 5.3. There are phase changes in the harmonics related to D1 at full load. The harmonics

Table 5.3.: Diagonal and Time Harmonics D1 pole coverage 0.6, SO 5*mm*.

Diagonal	Br	Br Phase	Bt	Bt Phase	S1 Time	Br	Br Phase	Bt	Bt Phase
S1T1	0.7424	102	0.0668	-55.4					
S3T3	0.1	-93.5	0.0233	-163.2	S3T1	0.0003	85.8	0.00003	-124
S5T5	0.147	-90	0.03	178.7	S5T1	0.0277	-51.5	0.0265	-141.8
S7T7	0.0144	-79.5	0.0032	86.4	S7T1	0.0182	-106	0.0176	-15.8
S9T9	0.0539	90	0.0184	2	S9T1	0.0002	93.2	0.0001	-26.2
S11T11	0.0247	92.4	0.0087	-8.4	S11T1	0.0255	-10.7	0.0254	-100.5
S13T13	0.0177	-91	0.008	-176.2	S13T1	0.0257	-145	0.0256	-55.1
S17T17	0.0024	86.8	0.0013	11.8	S17T1	0.0218	-50	0.0221	-140.4
S19T19	0.0118	90.2	0.0068	0	S19T1	0.0201	-102.7	0.0205	-12.6

pairs $r - S5T5$, $t - S5T1$ and $r - S7T7$, $t - S7T1$ are taking care of the magnitude as shown in Figure 5.17(b). The other pairs $r - S11T11$, $t - S11T1$ and $r - S13T13$, $t - S13T1$ are changing the phase of the curve which is shown in Figure 5.17(c). The combinations $r - S17T17$, $t - S17T1$ and $r - S19T19$, $t - S19T1$ make small changes in the peak values. It is evident that when all the harmonics are included, the curve obtained from FEM matches the reconstructed curve (Figure 5.17(a)).

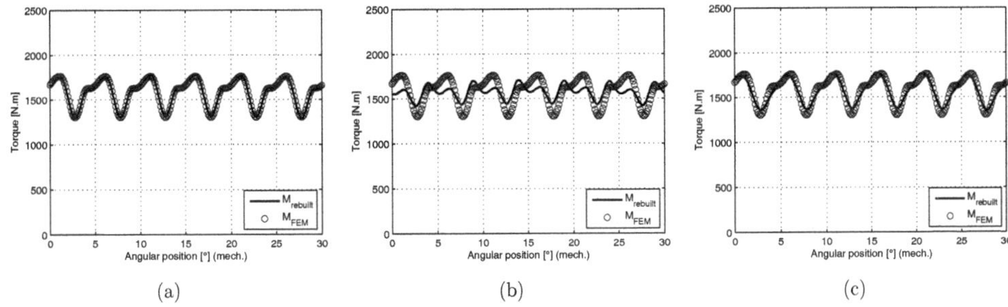

Figure 5.17.: Torque ripple D1 pole coverage 0.6, SO $5mm$, $I = I_N/2$. Torque ripple (a)Basic components. (b)Without $5, 7$ (c)Without $11, 13$.

Half load analysis for pole coverage at 0.7:

Table 5.4 lists the harmonics that generate torque ripple for pole coverage 0.7. Contrary to the pole coverage 0.6, torque ripple are not making any changes without harmonics order 17th and 19th.

Table 5.4.: Diagonal and Time Harmonics D1 pole coverage 0.7, SO $5mm$.

Diagonal	Br	Br Phase	Bt	Bt Phase	S1 Time	Br	Br Phase	Bt	Bt Phase
S1T1	0.7865	102.5	0.0669	-53.3					
S5T5	0.1156	-90.7	0.0265	-175.4	S5T1	0.0303	-55.5	0.028	-145.7
S7T7	0.084	-88.5	0.021	174.6	S7T1	0.0176	103	0.0165	13
S11T11	0.0292	89.4	0.0116	3	S11T1	0.0255	-12.2	0.0254	-101.9
S13T13	0.0278	91	0.0119	-2.4	S13T1	0.0262	-143.3	0.0263	-52.9

Figure 5.18 (a) shows that torque can be reconstructed from the harmonics pairs $r - S5T5$, $t - S5T1$, $r - S7T7$, $t - S7T1$, $r - S11T11$, $t - S11T1$ and $r - S13T13$, $t - S13T1$. Torque ripple can be eliminated if there are no harmonics order of $r - S5T5$, $t - S5T1$ and $r - S7T7$, $t - S7T1$. This is evident from Figure 5.18(b). Harmonics order of $r - S11T11$, $t - S11T1$ and $r - S13T13$, $t - S13T1$ which changes the phase is shown in Figure 5.18(c).

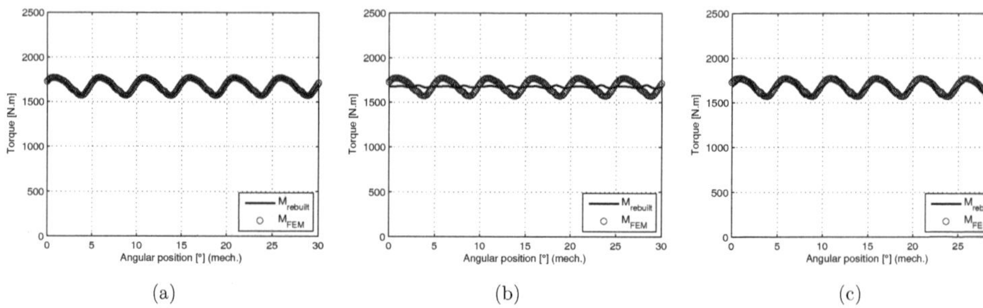

   (a)   (b)   (c)

Figure 5.18.: Torque ripple D1 pole coverage 0.7, SO $5mm$, $I = I_N/2$. Torque ripple (a)Basic components. (b)Without $5, 7$ (c)Without $11, 13$.

5.5.2. Torque Analysis at Half Load D3

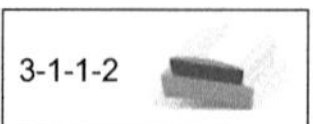

Half load analysis at D3 has additional harmonics that generate complete torque pattern similar to FEM. The harmonics for pole coverage 0.7 are listed in Table 5.5. The additional harmonics when compared to D1 are 3rd and 4th.

Table 5.5.: Diagonal and Time Harmonics D3 pole coverage 0.7, SO $5mm$.

Diagonal	Br	Br Phase	Bt	Bt Phase	S1 Time	Br	Br Phase	Bt	Bt Phase
S1T1	0.8214	104.7	0.0661	-49.9					
S2T5	0.0160	109	0.0009	78.6	S2T1	0.1493	-33.1	0.0465	-116.7
S3T3	0.0817	22	0.0224	-104.5	S3T1	0.0005	11.5	0.0001	-93.7
S4T7	0.0057	124.3	0.0025	71.6	S4T1	0.0969	-132	0.0297	-13.8
S5T5	0.0369	-50.2	0.01	-151.6	S5T1	0.0443	-51.5	0.0265	-141.8
S7T7	0.0376	-89.8	0.0103	165.2	S7T1	0.0177	-115	0.0075	4.8
S11T11	0.0044	-143.3	0.002	98.8	S11T1	0.0301	-13.5	0.0235	-100.5
S13T13	0.0048	89.3	0.0024	-2.2	S13T1	0.0289	-140	0.0234	-47

In Figure 5.19(a), the curves from FEM and the reconstructed curve are in good agreement. If harmonics order $r - S5T5$, $t - S5T1$ and $r - S7T7$, $t - S7T1$ are neglected, the reconstructed curve (Figure 5.19(c)) appears to be of same magnitude only with slight change in the phase. In this case, 5th and 7th are not making any significant change in torque ripple. However, $r - S2T5$, $t - S2T1$ and $r - S4T7$, $t - S4T1$ lessens the magnitude and change the phase of the curve (Figure 5.19(b)).

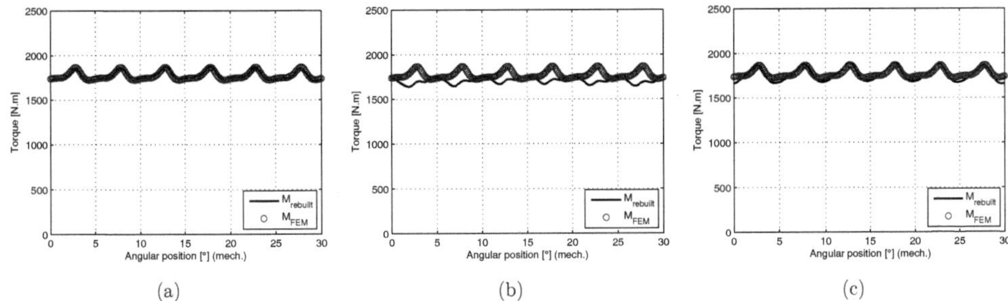

Figure 5.19.: Torque ripple D3 pole coverage 0.7, SO $5mm$, $I = I_N/2$. Torque ripple (a)Basic components. (b)Without $2, 4$ (c)Without $5, 7$.

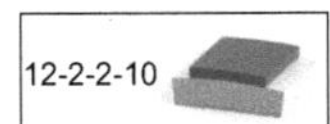

5.5.3. Torque Analysis at Half Load D4

Half load analysis for pole coverage at 0.6:

Similar to D4 under full load, the half load harmonics at pole coverage 0.6 is tabulated in Table 5.6. Only three dominant harmonics that generate torque pattern are provided in the Table 5.6.

Table 5.6.: Diagonal and Time Harmonics D4 pole coverage 0.6, SO $5mm$.

Diagonal	Br	Br Phase	Bt	Bt Phase	S1 Time	Br	Br Phase	Bt	Bt Phase
S5T1	0.6973	100.7	0.0839	-41.8					
S55T11	0.0197	91.6	0.0088	-48.7	S55T1	0.0241	-20.6	0.0241	-111
S65T13	0.0129	-90.3	0.0068	-178.3	S65T1	0.0313	-119	0.0315	-29

The harmonics pairs $r - S55T1$, $t - S55T1$ and $Sr - 65T13$, $t - 65T1$ are the working harmonics for torque pulsation of D4 for pole coverage 0.6. The two curves that are obtained through FEM and the reconstruction are depicted in Figure 5.20(a). Figure 5.20(b) shows nearly a constant torque if 55th and 65th are not considered.

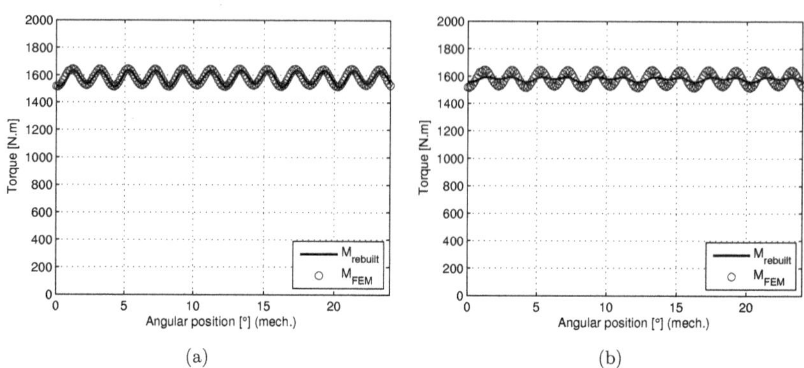

Figure 5.20.: Torque ripple D4 pole coverage 0.6, SO $5mm$, $I = I_N/2$. Torque ripple (a)Basic components. (b)Without $55, 75$

Half load analysis for pole coverage at 0.7:

Table 5.7 lists the harmonics for D4 arrangement at pole coverage 0.7. In this pole coverage, lot of harmonics pairs will generate torque ripple at half load.

Table 5.7.: Diagonal and Time Harmonics D4 pole coverage 0.7, SO $5mm$.

Diagonal	Br	Br Phase	Bt	Bt Phase	S1 Time	Br	Br Phase	Bt	Bt Phase
S5T1	0.7371	101	0.0846	-40					
S25T5	0.1041	-91.6	0.0281	-172.5	S25T1	0.0031	122.3	0.0025	38
S35T7	0.0723	-89	0.0223	176.2	S35T1	0.0014	-154.4	0.0012	-87.6
S55T11	0.0226	89.5	0.0103	2.2	S55T1	0.025	-23.3	0.025	-113
S65T13	0.0212	90.3	0.011	0	S65T1	0.032	-117	0.0325	-26.8

The torque ripple obtained from FEM and the torque ripple through reconstruction are shown in Figure 5.21. It is clear from Figure 5.21 that both the curves are in good agreement.

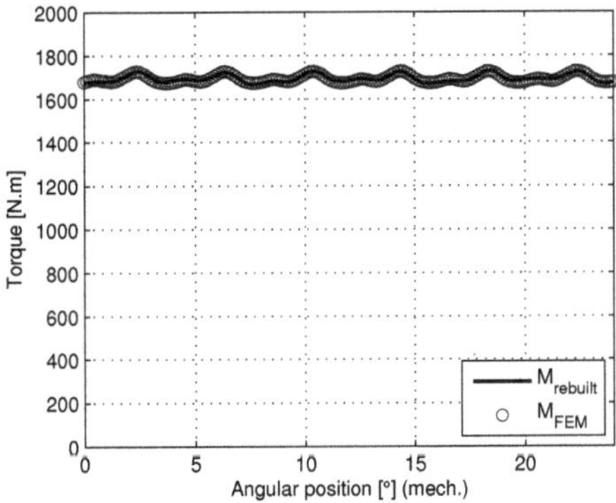

Figure 5.21.: Torque ripple D4 pole coverage 0.7, SO $5mm$, $I = I_N/2$. Torque ripple.

5.5.4. Torque Analysis at Half Load D6

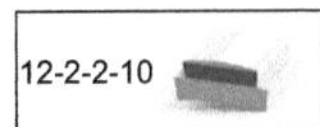

The list of harmonics at the air gap that create torque pulsation for D6 at half load for pole coverage 0.7 is shown in Table 5.8. Similar to the full load condition, 17th harmonics appears. As explained for geometry D6 under full load, the 17th harmonics for geometry D6 in half load

Table 5.8.: Diagonal and Time Harmonics D6 pole coverage 0.7, SO $5mm$.

Diagonal	Br	Br Phase	Bt	Bt Phase	1S Time	Br	Br Phase	Bt	Bt Phase
S5T1	0.7216	103.7	0.0823	-38.5					
S17T1	0.1062	-134.5	0.042	-9					
S25T5	0.0349	-75.2	0.0121	-165.2	S25T1	0.0025	107.8	0.0011	24.5
S35T7	0.0304	-99	0.011	161.4	S35T1	0.0015	153.3	0.0009	-107.5
S55T11	0.0035	94.6	0.0017	7.7	S55T1	0.0259	-21.5	0.0226	-108.4
S65T13	0.0048	75.8	0.0024	-16.5	S65T1	0.0298	-112.5	0.0273	-20

also makes DC shift in the average torque. This effect is shown in Figure 5.22(b). Figure 5.22(a) is again a constant torque line and it is evident that the reconstruction of harmonics from the selected order matches exactly the torque ripple from FEM.

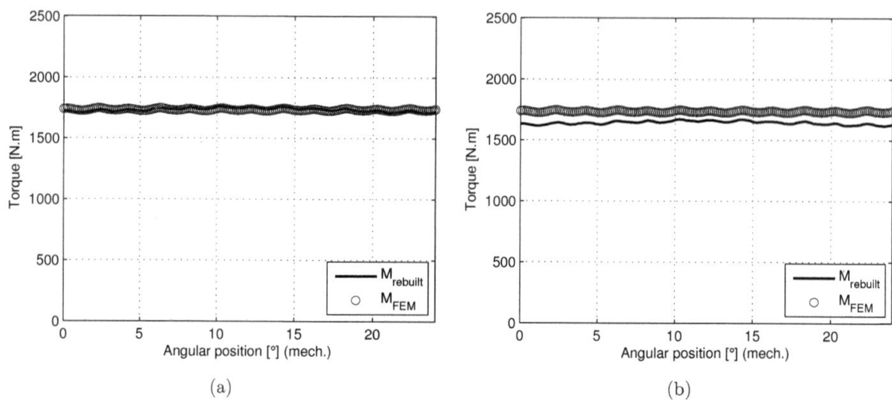

Figure 5.22.: Torque ripple D6 pole coverage 0.7, SO $5mm$, $I = I_N/2$. a)Basic components. (b)Without 17.

5.6. Summary

From the special study performed in this chapter, the following points are summarized:

- Using optimization methods, it is possible to reduce the torque pulsation by modifying the geometry. The dominant pulsating torque are again generated from harmonic pairs 5th and 7th. After optimization, D1 geometry has reduced torque pulsation because the magnitude of the harmonics is reduced.

- In the case of D3 geometry, the torque pulsation magnitude is not influenced by optimization algorithm. This is because there is no reduction in magnitude of harmonic pairs 5th and 7th in D3 structure after optimization. Moreover, the THD factor is less as shown in the Table 3.9 for D3 structure.

- Under field weakening operating mode, the torque ripple is increased from the harmonic pairs 5th and 7th. The harmonic pairs 11th and 13th are responsible for the phase change in the curve (see Figure 5.11)

- As discussed in section 5.4, axial shift of rotor reduces the torque ripple if there is room to shift the rotor.

- The half load analysis is performed to study the influence of harmonics due to saturation. The harmonic pairs responsible for the generation of the torque ripple that appear in the full load condition as studied in the sections (4.3, 4.5, 4.6, 4.8) also appear in the half load condition but with reduced magnitude of flux density components. Thus during half load conditions the magnitude of the torque ripple is reduced. There is no generation of new flux density harmonics pattern.

6. Generic Design Rules for Tooth Coil Winding Arrangement

Existing design rules in general are formulated based on several parameters especially to mitigate the torque pulsation for PMSM. In winding arrangement types such as Tooth-coil winding of PMSM, there are several harmonics interacting in the air gap that may generate torque pulsation. Also, there are several geometrical parameters that are influencing the torque pulsation. In this work, as examples six geometrical combinations and a measured result are selected for systematic investigation of parasitic torque. Now, it is possible to derive a common rule to select appropriate torque pulsation elimination methods by looking at harmonic-waves in the air gap field.

6.1. Design Guidelines

As pointed out in Chapter 1, the work aims at providing generic design guidelines for PMSM with tooth coil winding arrangement in terms of pulsating torque for Low Speed Applications. The pulsating torque mitigation is not limited to the tooth coils winding arrangement for PMSM. Certainly, it can be applied to other winding types (integer or fractional slot type with $q > 0$) and also for rotor type with electrical excitation. The design rules in this work developed for PMSM with Tooth coil winding arrangement are classified as follows.

- Design rules based on geometrical parameters.

- Simplified design rules.

- Design rules from manufacturing view point.

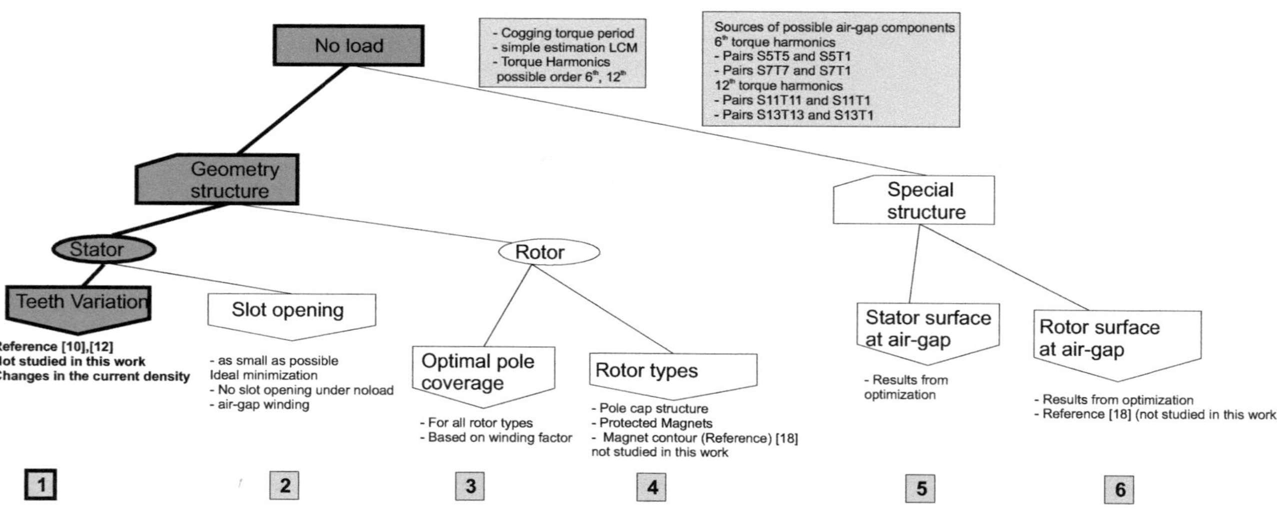

Figure 6.1.: Structure Example.

An example structure path of the design rules is shown in Figure 6.1. In this figure, each component in the electrical machine magnetic circuit is divided and organized as path to consider the design rules. In section 6.1.1, if it is highlighted "According to 1", it denotes the path 1 (please see Figure 6.1). Thus the numbers denote the possible path to select design rules to attain minimum torque pulsation.

The design rules have been divided into two main structures, one under no-load operation and the other under load operation. Figure 6.2 shows the structure under no-load. The paths from 1 to 6 in no-load structure are for geometry modification such as teeth, slot opening, pole coverage, rotor types and special structure. The structure of design rules for load operation is depicted in Figure 6.3. In load structure additional to the no-load structure paths 1-6, paths from 7 to 11 for stator winding arrangements and field weakening operation are considered.

6.1.1. Design Rules based on Geometrical Parameters.

6.1.1.1. No-load Case.

At no-load, the main parameters influencing the torque pulsation are the slot-opening and the pole coverage. The LCM between number of slots and the poles gives idea about the torque periods. The possible harmonics pair sources for torque pulsation independent of the machine configuration are,

- If 6th torque order is conspicuous, then the pairs that generate cogging are $r - S5T5, t - S5T1$ and $r - S7T7, t - S7T1$.

- If 12th torque order is conspicuous, then the pairs that generate cogging are $r - S11T11$, $t - S11T1$ and $r - S13T13, t - S13T1$.

- Beyond 12th order, the torque amplitude is very less.

- Other harmonics pairs, for example $r - S4T4, t - S4T10$ will generate harmonics but the dominant are from the first two of above mentioned points.

From the study made in the chapter 4 and chapter 5, as a hint the harmonics magnitude of approximately above 0.12 T are dominant to generate torque pulsation if it find its torque producing combination pairs. Possible solutions to mitigate the above harmonics orders are,

- According to 2, selecting SO which has less harmonics magnitude .

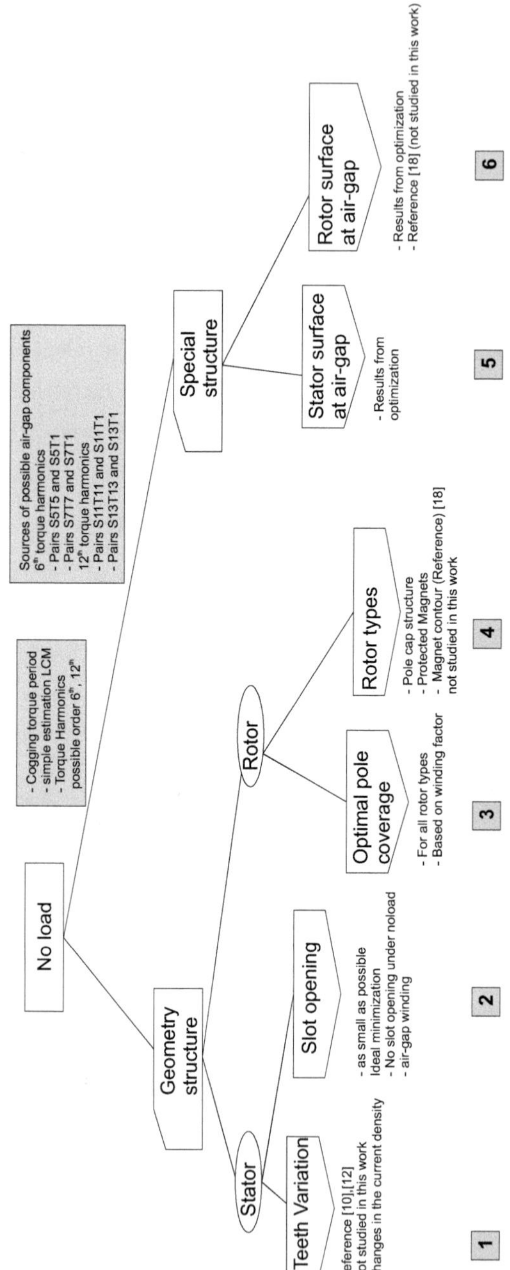

Figure 6.2.: No-load Structure.

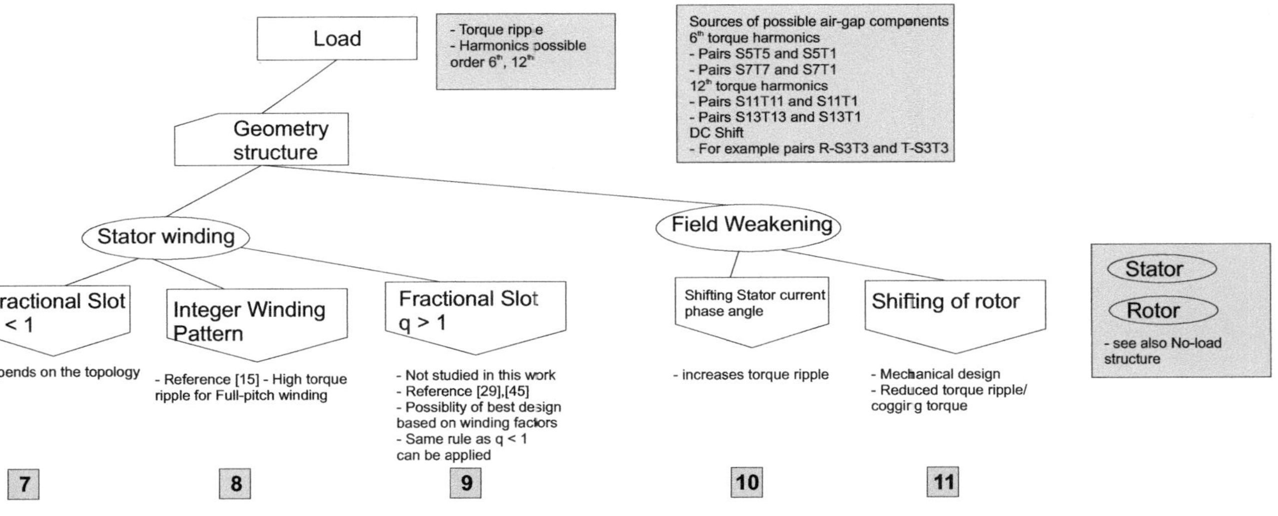

Figure 6.3.: Load Structure.

- According to 3, selecting pole coverage that has less harmonics magnitude.

- According to 4, rotor arrangement is selected which produces less harmonics amplitude.

- According to 5 and 6, the surface close to the air gap is altered to generate less harmonics magnitude. These modifications are performed using optimization algorithms.

- Torque harmonics can be shifted to higher frequencies by selecting elementary machines with high number of pole pairs and higher order of working wave, for example D4/D5/D6 in this work. Care should be taken, when selecting such configuration because it should not have poor winding factor and it should not generate radial force and noise.

- According to 1, there is possibility to have good machine with reduced cogging torque ([10], [12]).

6.1.1.2. Load Case

At load, the main geometrical parameters to influence the torque ripple are winding configuration, slot opening and pole coverage. Contrary to the no-load case, the periodicity of torque cannot be determined from the LCM between number of slots and poles. The sources of harmonics that generate torque ripple are as follows (remember that the sources are combined effect from stator and rotor fields),

- Harmonics pairs $r - S1T1$ from radial component and $t - S1T1$ from tangential component is the fundamental DC component.

- If 6th torque order is conspicuous, then the pairs that generate torque ripple are $r - S5T5, t - S5T1$ and $r - S7T7, t - S7T1$.

- If 12th torque order is conspicuous, then the pairs that generate torque ripple are $Sr - 11T11, t - S11T1$ and $r - S13T13, t - S13T1$.

- Beyond 12th order the torque amplitude is very less. Some configurations D4/D5/D6 produce sub-harmonics torque whose amplitude is very less.

- Other harmonics pairs, for example $r - S3T1$ from radial component and $t - S3T1$ from tangential component generate DC shift, this is due to the saturation effects as discussed in Chapter 4 (depends on the pole coverage either positive or negative). If there is further increase in saturation, the small amount of constant torque will have less influence in the additional torque because the fundamental harmonic which is responsible for fundamental constant torque will be reduced thus reducing the total torque.

The possible solutions to mitigate the above harmonics orders at load are,

- According to 2, selecting SO which has less harmonics magnitude .

- According to 3, selecting pole coverage that has less harmonics magnitude. Depending on the geometry arrangement, there can be 2 or 3 or more minimum pole coverage points.

- According to 4, rotor arrangement is selected which produce less harmonics amplitude.

- According to 5 and 6, the surface close to the air gap is altered to generate less harmonics magnitude. These modifications are performed using optimization algorithms.

- According to 7, it is wise to select winding topology for low speed machines. Torque harmonics can be shifted to higher frequencies by selecting high pole pairs, for example D4/D5/D6 in this work. Care should be taken when selecting such configuration because it should not have poor winding factor and it should not generate radial force and noise.

- According to 1, there is possibility to have good machine with reduced cogging torque ([10], [12]).

- According to 8, full-pitch integer winding has high torque pulsation due to high magnitude of harmonics [15]. It is not recommended for the low-speed machines. But integer winding with short-pitch arrangement [43] and [29], for e.g. integer winding $q = 2$ with short-pitch $W/\tau_p = 5/6$, will have reduced torque ripple due to combination pairs $r - S5T5, t - S5T1$ and r $S7T7, t$ $S7T1$. This type of winding is not studied in the work.

- According to 9, fractional slot winding type [29] of $q > 1$ will have reduced torque ripple due to combination pairs $r - S5T5, t - S5T1$ and $r - S7T7, t - S7T1$. This configuration is not studied in this work.

- According to 10, torque ripple is increased during field weakening operation as discussed in Section 5.3. Torque ripple can be reduced according to 11.

- According to 11, due to axial shifting of rotor, the harmonics magnitude is less. This is studied in Section 5.4.

6.1.2. Simplified Design Rules

Based on the study from previous chapter, simplified design rules can be formulated for minimum pulsating torque. The following points should be remembered,

- As discussed in chapter 3, the geometry which has less cogging torque has less torque ripple. The question is what could be the magnitude of cogging torque? The answer is according to the paths 2,3 and 4 (please refer to Figure 6.2) in the no-load structure, selecting less magnitude of harmonics flux density, for example, harmonics flux density value less than 0.1 T.

- As discussed in chapter 3, section 3.3.3, the pole coverage can be selected according to the winding factor. The Equation 3.96 is zero when α_p $(2/5, 4/5, 8/5, \ldots)$ is selected. Thus torque pulsation due to 5th harmonics can be reduced with pole coverage of 4/5. If α_p selected as $(2/7, 4/7, 8/7, \ldots)$ then torque pulsation due to 7th harmonics can be reduced thus eliminating its effect. This effect in the design rules is according to the path 7,8 and 9 (please refer to Figure 6.3) in the load structure.

- According to 4, the rotor arrangement which has less harmonics amplitude is selected.

6.1.3. Design Rules from Manufacturing View Point

Sometimes, the production cost is high although the machine produces less torque pulsation, for example, contour magnets configuration. In this section, some rules related to the manufacturing aspects are provided.

- According to 4, the rectangular magnets are easy to manufacture and this reduces the production costs. Pole-cap configuration is a possible solution, which has less harmonics magnitude.

- According to 7,8 and 9, configuration 7 is a tooth coil winding arrangement. In design rules path 7, the winding can be wound independent of the stator lamination. Later, it can be inserted into the slot, which reduces the time during production. Thus the machine deliverable time is faster. This winding configuration has less harmonics in the air gap flux density, according to the study made in Chapter 4.

- According to 7, the maintenance is easy, if there are problems later with stator winding failure. This reduces the maintenance cost.

- According to 7, the winding is suitable for low-speed application because it has increased pole pairs. Thus, stator yoke can be reduced if there is enough mechanical stiffness in the yoke. Thus, material costs can be reduced.

- According to 11, the rotor shifting arrangement can be used to reduce the inverter costs, if there is room for axial shift and if it is mechanically feasible. This configuration has less harmonics magnitude as discussed Section 5.4.

7. Conclusion and Outlook

The main aim of this thesis work is to recommend the design rules for Permanent Magnet Synchronous Machine with tooth coil winding arrangement for low-speed applications. In the last decade, the industries started thinking about replacing the existing drive systems with gearbox by direct drives, which has no gear system. At the beginning of this work, selection of PMSM for low-speed applications is discussed. Initially, an extensive literature research is performed. This indicates the existing design rules for minimum pulsating torque. The existing rules have no methodological approach in reducing the parasitic torques by considering harmonics in the air gap field. However, in this work, the generic design rules are established systematically by considering the harmonics in the air gap field. As an example for a low speed application, a direct drive for paper mill is considered.

In this work, as examples six typical exemplary geometries are studied to derive the design rules for PMSM with minimized pulsating torque. In general, the sources for parasitic torque are harmonics from stator field, rotor field and permeance function. The field equations of stator and rotor fields are superimposed with permeance functions, yielding the total field in the air gap. From the air gap field information (radial and tangential components), torque is derived analytically based on Maxwell Stress Tensor method. The analytical result for surface mounted machine is compared with FEM.

Although the analytical calculation method is faster, there is limitation in this calculation due to second order effects. This work is proceeded using FEM software for the considered geometries. The air gap radial and tangential flux densities are obtained from numerical calculation for every space and time. The 2D geometries are used in FEM calculation for an elementary machine. With the data available from numerical calculation, 2D analysis, an elegant approach is carried out to analyze the reason for the parasitic torque. It is found that only harmonics of same pole pairs are the source for pulsating torque. It is clearly mentioned that only few combinations are responsible and also there are harmonics responsible for DC shift.

Experimental test bench was prepared for geometry D3 to validate the numerical calculation. The results from the measurement are in good agreement with the numerical calculation. Afterwards, the analysis is continued with a special study for problem such as optimization, field weakening and half load condition. In the special study, optimization is performed to modify the air gap surface on stator to obtain minimum cogging torque. Torque ripple minimization during field weakening is explained by mechanically shifting the rotor in the axial direction. Finally, torque ripple during half load operation is studied to check the influence of saturation. For all the studied work including measured results, the reason for pulsating torque is provided. It can be proved that no new pattern on torque harmonics is available.

Outlook

Finally, this thesis work concludes with the design rules for PMSM that gives an electrical machine designer an easy to understand insight into the source of pulsating torque in the machine. It is also mentioned that the 2D approach on design rules is also valid for other winding arrangements and also for electrically excited synchronous machine.

This work can be continued by extending the analysis to the problems such as eccentricity as well as material defects and their influence on pulsating torque. The influence of lamination materials and magnet materials can also be studied. This study can be extended to the noise and vibration problems of synchronous machines. Further, the impact of inverter supply on the generation of new patterns of torque ripple can also be investigated. The boundary condition set for the application in this work is for low speed application. It can also be studied for high speed machines. Influence of geometrical parameters such as air gap variation, outer diameter of stator variation, inner diameter of rotor variation, teeth variation, slot variation, yoke variation, flux barrier, etc. can be investigated. In all the cases, it should be kept in mind about the cost of production and maintenance of the machine. The design rules can also be combined with the existing design rules discussed in Chapter 1 to deliver a good solution for the electrical machine designers.

A. Schwartz-Christoffel Transformation

Schwartz Christoffel tranformation maps the interior of the polygon of w-plane into the upper half of the z-plane as illustrated in Figure A.1 [73]. The Equation for this tranformation is,

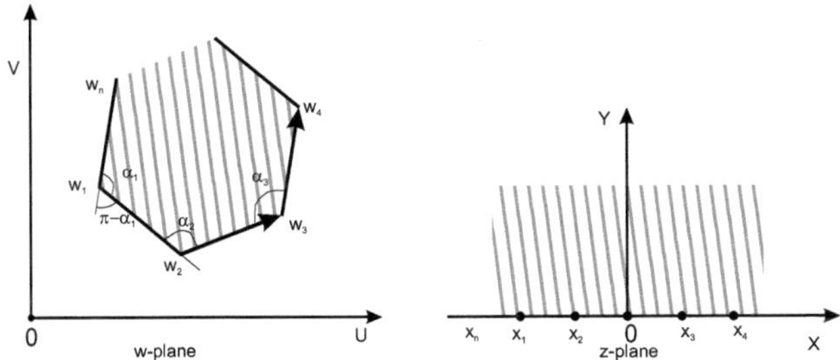

Figure A.1.: Illustration of a Schwartz Christoffel mapping.

$$\frac{dw}{dz} = A(z - x_1)^{\frac{\alpha_1}{\pi} - 1} \cdot (z - x_2)^{\frac{\alpha_2}{\pi} - 1} \dots (z - x_n)^{\frac{\alpha_n}{\pi} - 1} \tag{A.1}$$

where α_n are the interior angle of the polygon and w_n are the vertices as shown in Figure A.1. These vertices are mapped into vertices x_n in the z-plane.

$$w = C_1 \int (z - x_1)^{\frac{\alpha_1}{\pi} - 1} \cdot (z - x_2)^{\frac{\alpha_2}{\pi} - 1} \dots (z - x_n)^{\frac{\alpha_n}{\pi} - 1} + C_2 \tag{A.2}$$

where the constants determine the size and position of polygon. When w-plane moves from $w - n$ towards $w - 1$, the z-plane moves from x_n to x_1. The corresponding angle in the z-plane changes as well. This process proceeds and all the vertices are mapped from w-plane to z-plane.

B. Field equation of current sheet

The field solution for current sheet using vector potential is explained in cylindrical co-ordinates in [74] and [43]. The cylindrical model from [43] is derived here. The Laplace equation in cylindrical coordinates is,

$$\frac{\partial^2 V(r)}{\partial r^2} + \frac{1}{r}\frac{\partial^2 V(r)}{\partial r} + \frac{p^2}{r^2} V(r) = 0 \tag{B.1}$$

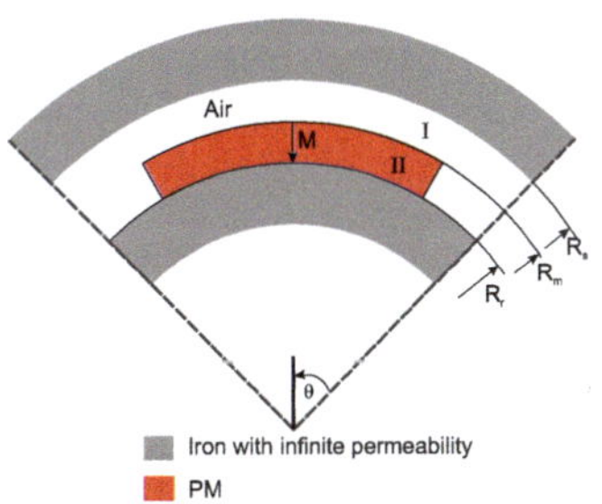

Figure B.1.: Geometry illustration for Stator field.

The general solution of the partial differential equation for the z-component is,

$$V(r) = \mu_0 \hat{A}_z \frac{r_i}{p} \left[C \left(\frac{r}{r_i} \right)^{p} + D \left(\frac{r}{r_i} \right)^{-p} \right] \tag{B.2}$$

the radial flux density and tangential flux density are,

$$B_r = -\frac{1}{r}\frac{\partial V}{\partial \theta} \tag{B.3}$$

$$B_\theta = \frac{\partial V}{\partial r} \tag{B.4}$$

The normal flux density distribution of the stator field is,

$$B_{sa}(r_a) = j\mu_0 \hat{A}_z \left[C_I \left(\frac{r_a}{r_i} \right)^{p-1} + D_I \left(\frac{r_a}{r_i} \right)^{-p-1} \right] \tag{B.5}$$

where the constants C_I and D_I are determined using boundary conditions and it is given in lecture script [43].

at $r = r_a$, $B_{\theta I} = -\mu_0 \cdot A_z$

at $r = r_m$, $B_{rI} = B_{rII}$

at $r = r_m$, $B_{\theta II} = \mu_{rM} B_{\theta I}$

at $r = r_i$, $B_{\theta II} = 0$

C. Modal analysis of rotors D2 and D3

The mechanical stress calculations for the iron ring rotor and pole-cap rotor are performed using ANSYS FEM program. Standard material properties such as yield strength and poisson's ratio from the datasheet are used for the calculations.

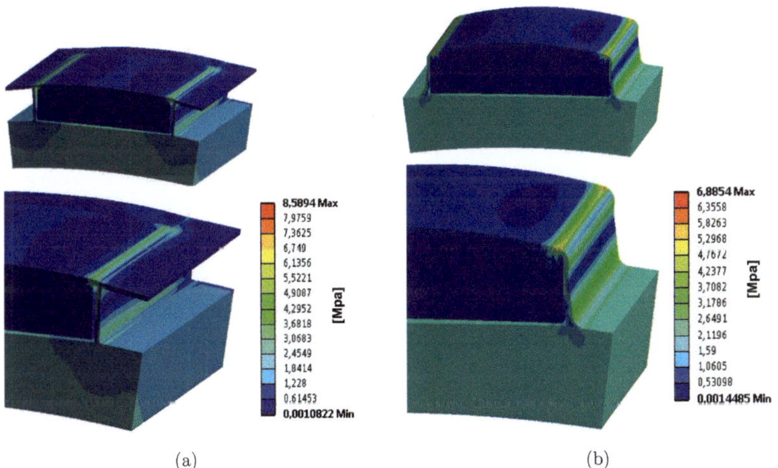

Figure C.1.: Model analysis.(a)D2 (b) D3

The typical value for iron sheet is 450 N/mm^2, when considering the safety factor it is 200 N/mm2. In the FEM, elasto-plastic material property is employed. The von-Mises-stress for 20% more than the maximum speed 400 min^{-1} is shown in Figure C.1. The calculated maximum von-Mises-stress are $8.6N/mm^2$ for iron ring rotor and $6.9N/mm^2$ for pole-cap rotor which are very less when compared to the available yield strength of the material.

D. FEM mesh and post processing figures

There are fine quadrilateral mesh in the air gap region as shown in Figure D.1. This type of mesh is employed because, cogging torque and torque ripple calculation will have good convergence. The flux density value is extracted exactly in the air-gap at middle of each mesh. This is because, the average value of potential lie at the middle of the mesh.

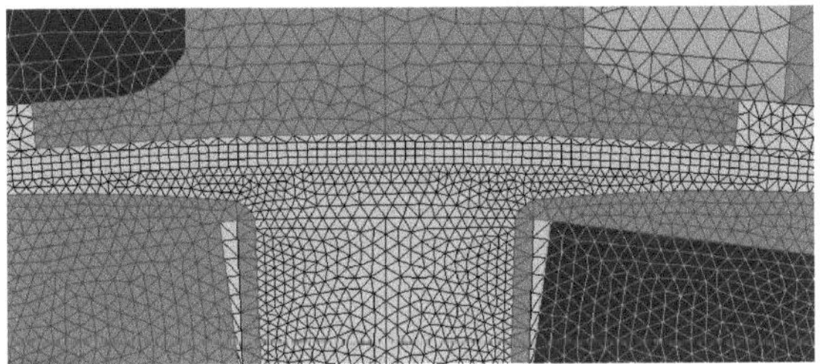

Figure D.1.: Air gap mesh.

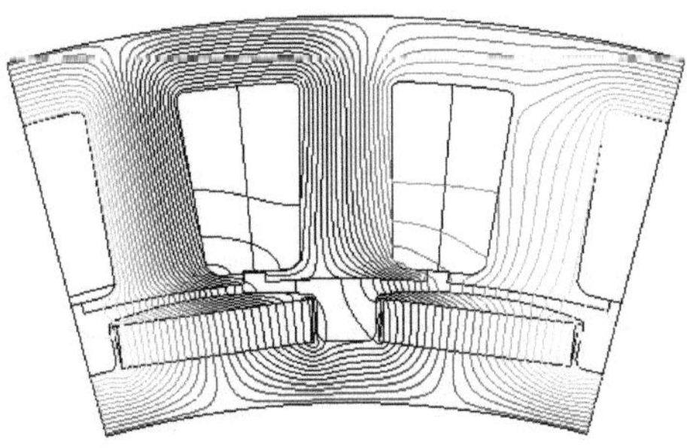

Figure D.2.: Flux lines at rated load for D3.

In Figure D.2, the flux lines for D3 geometry cross section is shown in for one-elementary machine at rated load.

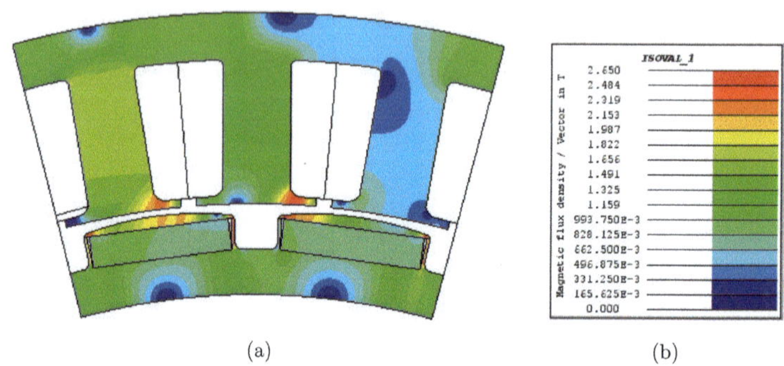

(a)   (b)

Figure D.3.: Flux density distribution at rated load for D3

In Figure D.3, the flux density distribution for D3 geometry cross section is shown in Figure for one-elementary machine at rated load. In this operating point, the flux density in the teeth is $1.66T$ and flux density at the yoke is $1.2T$, which is in the acceptable range. The teeth tip and pole-cap are saturated.

List of Figures

List of Tables

Symbols und Acronyms

α_p	Pole coverage (ratio of pole arc to pole pitch)
δ_m	Mechanical air gap
$\hat{A}_{s_1}$	Current sheet peak value
$\hat{B}_{\delta_1}$	No-load flux density peak value
$\lambda_{r,\theta}$	Air gap permeance function
φ	Phase angle
ρ_M	Volume charge model
ρ_{cu}	Copper material density
σ_{cu}	Conductivity of copper
τ_p	Pole pitch
τ_s	Slot pitch
τw	Tangential force density due to Lorentz force
τx	Tangential force density due to reluctance force
φ_I	Scalar potential of region I
φ_M	Scalar potential
$\vec{B}$	Magnetic flux density
$\vec{B}_I$	Flux density of region I
$\vec{B}_{II}$	Flux density of region II
$\vec{E}$	Electric field
$\vec{F}$	Lorentz Force
$\vec{H}_I$	Magnetic field intensity of region I
$\vec{H}_{II}$	Magnetic field intensity of region II
$\vec{J}$	Current density
$\vec{M}$	Magnetization of the permanent magnet

$\xi_{w\nu}$ .... Winding factor

ζ .... Ratio of slot opening to air gap length

A_1, A_2, B_a, B_2 .... Constants in PDE

A_s .... Stator current sheet

A_ν .... Current sheet harmonics

b_N .... Slot opening

B_r .... Flux density radial component

B_s .... Armature flux density

B_s, B_t, B_k .... Flux density in S-plane, T-plane, K-plane

B_ν .... Armature flux density harmonics

B_θ .... Flux density tangential component

$B_{r,\mu} B_{\theta,\nu}$ .... Flux densities radial and tangential with harmonics

B_{rotor} .... PM field

B_{stator} .... Armature field

B_{THD} .... No-load flux density total harmonic distortion factor

$B_{\delta q}$ .... Magnetic flux density including the slot effect

D .... Bore diameter

h_j .... Height of the stator yoke

h_m .... Height of PM

h_n .... Height of the teeth

I_d, I_q .... Current, d-axis and q-axis

i_g .... Transmission ratio

I_s .... Current effective value

J_{eff} .... Current density effective

k_c .... Carter factor

k_{sat} .... Saturation factor

L .... Active length

l .... length

M .... Torque

m .... Number of phases

M_N .... Rated torque

M_r .... Radial magnetization

$M_\%$ .... Relative cogging torque and torque ripple in percentage

M_Δ .... difference between maximum and minimum value of torque

M_θ .... Tangential magnetization

m_{active} .... Active mass

m_{cu} .... Mass of the copper

M_{out}, M_{in} .... Torque on load side and torque on drive side

n .... Speed

p .... pole pairs

P_{mech} .... Mechanical Power

$P_{v,cu}$ .... Loss of the copper

q .... Number of slots per pole per phase

Q_s .... Number of slots

r .... radius

R_m .... radius include PM

R_r .... inner radius until PM

R_s .... stator inner radius

r_{out}, r_{in} .... Radius of gear on load side and on drive side

R_{sa} .... Stator resistance

r_{st} .... Bore radius

s, z, w, t .... Subscript used in plane transformation

T .... Tensor

t .... time

U_d, U_q .... Voltage, d-axis and q-axis

w_s .... Number of turns

X_d, X_q .... Reactance, d-axis and q-axis

λ_{pm} .... Air gap permeance variation due to PM

λ_{sat} .... Permeance due to saturation

λ_s .... Slot permeance

μ_0 .... Permeability in free space

μ_{rM} .... Relative permeability

172

ω_m .... Angular speed

$\Omega_{min}, \Omega_{mout}$ .... Angular rotation on drive side and on load side

ψ_{PM} .... Flux linkage

$\rho_{average}$ .... Average material density

τ_w .... Force density

$2DFFT$ .... Two dimensional Fast Fourier Transform

AC .... Alternating Current

ASM .... Asynchronous machine

CO_2 .... Carbon dioxide

CT .... Conformal Transformation

DC .... Direct Current

FEM .... Finite Element Method

HCF .... Highest Common Factor

LCM .... Least Common Multiple

MA .... Maximum current

MF .... Maximum flux

$MTPA$ .... Maximum torque per ampere

$NdFeB$ .... Neodymium-Iron-Boron

PDE .... Partial Differential Equation

PM .... Permanent Magnet

$PMSM$ .... Permanent Magnet Synchronous Machine

SM .... Synchronous Machine

$SmCo$ .... Samarium-Cobalt

SO .... Slot Opening

THD .... Total Harmonic Distortion

Bibliography

[1] BMWi. Energy efficiency made in germany, 2008.

[2] Beste verfuegbare Techniken. Integrated pollution prevention and control (ippc) reference document on best available techniques in the pulp and paper industry. Technical report, BVT, 2001.

[3] VDP. Verband deutscher papierfabriken e. v., 2010.

[4] Scott.R. Semken, M. Polikarpova, P. Roytta, J. Alexandrova, J. Pyrhonen, J. Nerg, A. Mikkola, and J. Backman. *Direct-drive permanent magnet generators for high power wind turbines: benefits and limiting factors*. In the Institution of Engineering and Technology Renewable Power Generation, 2012. (IETLD '12), 2012.

[5] Lucian Nicolae Tutelea Ion Boldea. *Electric Machines: Steady State, Transients, and Design with MATLAB*. CRC Press, 2009.

[6] Jonas Georg. *Grundlagen zur Auslegung und Berechnung elektrischer Maschinen*. VDE, 2001.

[7] Sheppard J Salon. *Finite Element Analysis of Electrical Machines*. Springer, 2007.

[8] Pia Salminen. *Fractional slot permanent magnet synchronous motors for low speed applications*. PhD thesis, Lappeenranta University of Technology, 2004.

[9] T. Ishikawa and R.G. Slemon. *A method of reducing ripple torque in permanent magnet motors without skewing*. IEEE Transactions on Magnetics, 29(2), March 1993.

[10] Zhu Z.Q., David Howe. *Influence of design parameters on cogging torque in permanent magnet machines*. IEEE Transactions on Energy Conversions, 15, 2000.

[11] Zhu Z.Q. *Fractional slot permanent magnet brushless machines and drives for electric and hybrid propulsion systems.* International Conference on Ecologic Vehicles and Renewable Energies, 2009. (MONACO '09), 2009.

[12] Ackermann B., J.H.H. Janssen, R. Sottek, R.I. van Steen. *New technique for reducing cogging torque in a class of brushless dc motors.* Electric Power Applications, IEEE Proceedings B, 139(4):315 –320, jul 1992.

[13] Gueemes J. A., A. M. Iraolagoitia, P. Fernandez, M. P. Donsion. *Comparative study of pmsm with integer-slot and fractional-slot windings.* International Conference on Electrical Machines, 2010. (ROME '10), 2010.

[14] Dajaku G., D. Gerling: *Cost-effective and high-performace motor designs.* International Conference on Electric Drives Production, 2011. (EDPC '11), 2011.

[15] Magnussen F., P. Thelin, C. Sadarangani: *Performance evaluation of permanent magnet synchronous machines with concentrated and distributed windings including the effect of field-weakening.* IEEE Conference Publications, 2:679 – 685, 2004.

[16] Canders W. -R., H. Mosebach, Z. Shi: *Analytical and numerical investigation of pm-excited linear synchronous machines with shaped magnets.* International Symposium on Linear Drives for Industry Applications, 2003. (LDIA '03), 2003.

[17] Panu Kurronen: *Torque vibration model of axial-flux surface-mounted permanent magnet synchronous machine.* PhD thesis, Lappeenranta University of Technology, 2003.

[18] Krotsch J., T. Ley, B. Piepenbreier: *Reduction of torque and radial force fluctuation in permanent magnet synchronous motors by means of multi-objective optimization.* IEEE, pages 40 – 48, 2011.

[19] Gasparin L., A. Cernigoj, S. Markic, R. Fiser: *Additional cogging torque components in permanent-magnet motors due to manufacturing imperfections.* IEEE Transactions on Magnetics, 45(3):1210 – 1213, 2009.

[20] Coenen I., M. van der Giet, K. Hameyer: *Manufacturing tolerances: Estimation and prediction of cogging torque influenced by magnetization faults.* IEEE Transactions on Magnetics, 48(5):1932 – 1936, May 2012.

[21] Joerg Steinbrink: *Kraftwirkungen in permanentmagneterregten Maschinen.* PhD thesis, Leibniz University Hannover, 2006.

[22] Hafner M., D. Franck, K. Hameyer: *Accounting for saturation in conformal mapping modeling of a permanent magnet synchronous machine.* COMPEL, 30(3):916–928, May 2011.

[23] Tareilus G. P. Hoffmann, W.-R. Canders, H. Mosebach: *Large high performance linear drive with high overload capability and very small thrust ripple.* International Symposium on Linear Drives for Industry Applications, (LDIA '07), 2007.

[24] Ocak O., M. Aydin: *Influence of varying magnet pole-arcs and step-skew on permanent magnet ac synchronous motor performance.* IEEE Transactions on Magnetics, 2012.

[25] Hung J. Y., Z. Ding: *Design of currents to reduce torque ripple in brushless permanent magnet motors.* IEEE Proceedings B, 140(4):260 – 266, 1993.

[26] Petrovic V., R. Ortega, A.M. Stankovic, G. Tadmor: Design and implementation of an adaptive controller for torque ripple minimization in pm synchronous motors. IEEE Transactions on Power Electronics, 15(5):871 – 880, 2000.

[27] Canders W.-R., A.B. Asaf Ali: *High speed-parameterstudie.* Technical report, TU-Braunschweig, 2011.

[28] Peter Behrends: *Elektromaschinen und Antriebe.* Huethig and Pflaum Verlag, 2011.

[29] Pyrhoenen Juha., Tapani. Jokinen, Valeria. Hrabovcova: *Design of rotating electrical machines.* Wiley, 2008.

[30] Schaefer H.: *Wheel-hub drives vs. axle drives.* Symposium Hybridfahrzeuge und Energie-management,2009. (HEVEM '09), 2009.

[31] Thomas Finken, Kay Hameyer: *Design of electric motors for hybrid- and electric-vehicle applications.* 12th International Conference on Electrical Machines and Systems, ICEMS, Tokyo, Japan, November 2009.

[32] Lange A., Laube F., W.-R. Canders, H. Mosebach: *Comparison of different drive systems for a 75 kw electrical vehicle drive.* International Conference on Electrical Machines, (ICEM '00), 2000.

[33] Singh B., B.P. Singh, S. Dwivedi: *A state of art on different configurations of permanent magnet brushless machines.* IE(I) Journal EL - The Institution of Engineers (India), 87, 2006.

[34] Rik. De Doncker, Duco W.J. Pulle, Andre Veltman: *Advanced Electrical Drives*. Springer, 2011.

[35] Eduardo Burguete Archel: *Development of thermal model of pmsm for direct drive application*. Master's thesis, Technical University Braunschweig, 2010.

[36] VACUUMSCHMELZE: VACUUMSCHMELZE GmbH Co. KG, 2013.

[37] Mosebach.H.:   *Yearly Report*.   Systematik dreisträngiger symmetrischer PM-erregter PPSM, 2005.

[38] Canders W.-R., D. Huelsmann: *Analysis and determination of symmetrical three phase windings with focus on tooth coil windings*. International Symposium on Electromagnetical Fields, (ISEF '11), 2011.

[39] Nicola. Bianchi, Massimo. Barcaro, Silverio. Bolognani: *Electromagnetic and Thermal Analysis of Permanent Magnet Synchronous Machines*. InTech, 2012.

[40] Cistelecan, M.V., Ferreira, F.J.T.E., M. Popescu: *Three phase tooth-concentrated multiple-layer fractional windings with low space harmonic content*. Energy Conversion Congress and Exposition (ECCE), IEEE, 1399-1405, Sept 2010.

[41] Zhu Z.Q., David Howe: *Analytical prediction of the cogging torque in radial-field permanent magnet brushless motor*. IEEE TRANSACTIONS ON MAGNETICS, 28(2), March 1992.

[42] Edward P. Furlani: *Permanent Magnet and Electromechanical Devices*. Elsevier, 2001.

[43] Canders W.-R.: *Energieumformung i*. Technical report, Institut für Elektrische Maschinen, Antriebe und Bahnen: TU-Braunschweig, 2010/11.

[44] Tingley E.M.: *Two-Phase and Three-Phase Lap Windings in unequal Groups*. The Electrical Review and Western Electrician,Vol.66, No.4,S.166 , 1915.

[45] Vogt K.: *Berechnung elektrischer Maschinen*. VCH Verlag weinheim, 1996.

[46] Binder A., M. Mirzaei, B. Funieru: *Acoustic noise calculation for high-speed permanent-magnet motors*. ELECTROMOTION 2009 EPE Chapter Electric Drives Joint Symposium, July 2009.

[47] Binns K.J., P.J. Lawrenson: *Analysis and Computation of Electric and Magnetic Field Problems*. Pergamon Press, 1963.

[48] Drago Ban, Damir Zarko, Thomas A. Lipo: *Analytical calculation of magnetic field distribution in the slotted air gap of a surface permanent-magnet motor using complex relative air-gap permeance.* IEEE TRANSACTIONS ON MAGNETICS, 42(7), July 2006.

[49] Drago Ban, Damir Zarko, Thomas A. Lipo: *Analytical solution for electromagnetic torque in surface permanent-magnet motors using conformal mapping.* IEEE TRANSACTIONS ON MAGNETICS, 45(7), 2009.

[50] Seinsch H.O.: *Oberfelderscheinungen in Drehfeldmaschinen.* Teubner Verlag Stuttgart, 1992.

[51] Bedrich Heller, Vaclav Hamata: *Harmonic Field Effects in induction Machine.* Academia Publishing House, 1977.

[52] Karl Vogt Germar Müller, Bernd Ponick: *Berechnung elektrischer Maschinen.* John Wiley Sons, Febraury 2012.

[53] Jimoh A. A., R.D. Findlay: *Parasitic torques in saturated induction motors.* IEEE Transactions on Energy Conversion, 3(1), March 1988.

[54] Hassanzadeh Kiyoumarsi A. M. R., M. Moallem: *A new analytical method on the field calculation of interior permanent-magnet synchronous motors.* IEEE, volume 13, pages 364–372, October 2006.

[55] Timar P. L., Laszlo Timar-Peregrin Timar-P: *Noise and vibration of electrical machines.* Elsevier, 1989.

[56] Annette Muetze Ewgenij Starschich, Kay Hameyer: *Analytical force calculation in brushless-dc motors i: An alternative approach and analytical force calculation in brushless-dc motors ii: Mathematical details of the alternative approach.* Electric Machines and Drives Conference, IEMDC, IEEE International, pages 1200–1207, 2009.

[57] Prof. Gerhard Henneberger: *Electrical Machines I.* Technical report, Institut für Elektrische Maschinen: RWTH-Aachen, 2001.

[58] Canders W.-R., A. Asaf Ali: *Torque ripple situation in high torque synchronous machine with protected magnets.* International Symposium on Electromagnetical Fields, (ISEF '09), 2009.

[59] Miyamoto Y., T.Higuchi, T.Abe: *Consideration for fractional slot winding of permanent magnet type synchronous machine.* International Conference on Electrical Machines and Systems, ICEMS, pages 1-6, Beijing , China, 2011.

[60] CEDRAT: *Cedrat design solutions for electrical engineering,* 2010.

[61] Reichert K.: *A simplified approach to permanent magnet and reluctance motor characteristics determination by finite-element methods.* COMPEL. 2006, 25, 368-378.

[62] Lukaniszyn M., M. Jagiela, R. Wrobel, K. Latawiec: *2D harmonic analysis of the cogging torque in synchronous permanent magnet machines* COMPEL: The International Journal for Computation and Mathematics in Electrical and Electronic Engineering, Vol. 23 Iss: 3, pp.774 - 782.

[63] Rothe R., M. van der Giet, K. Hameyer: *Convolution approach for analysis of magnetic forces in electrical machines.* COMPEL. 2010, 29, 6.

[64] Salon S., K. Sivasubramaniam, L.T. Ergene: *The effect of asymmetry on torque in permanent magnet motors.* 2000, 28-30. ICEM.

[65] Salon S., K. Sivasubramaniam, L.T. Ergene: *Finite Element Analysis of Electrical Machines.* 2001, 208-217. IEMDC.

[66] Van der Giet M., R. Rothe, M.H. Gracia, K. Hameyer: *Analysis of noise exciting magnetic force waves by means of numerical simulation and a space vector definition.* 2008, 1-6. ICEM.

[67] Ahamed Bilal A.: *Annual Report.* Tools Capable of Performing Optimization for Electrical Machines, Institut für Elektrische Maschinen, Antriebe und Bahnen: TU-Braunschweig, 2010/11.

[68] FGOT. FGOT, Prof. Jean-Louis Coulomb, Grenoble Electrical Engineering Laboratory, G2ELAB, Grenoble-INP, Grenoble, France, 2012.

[69] Schlensok C., S. Felgentraeger, D. van Riesen, K. Hameyer: *Stator-Tooth Optimization of an Induction Machine with Squirrel-Cage Rotor.* 2006. ICEM.

[70] Xin Cong: *Berechnung von verschiedenen Betriebspunkten einer Permanent Magnet Erregte Synchronmaschine.* Master Arbeit, Technical University Braunschweig, 2011.

[71] Lipo T.A., M. Aydin: *Field Weakening of Permanent Magnet Machines Design Approaches* EPE Power Electronics and Motion Control Conference, EPE-PEMC'04, Sept. 2004, Riga, Latvia.

[72] Henning Woehl-Bruhn: *Synchronmaschine mit eingebetteten Magneten und neuartiger variabler Erregung für Hybridantriebe.* PhD thesis, Technical University Braunschweig, 2010.

[73] Grewal B.S: *Higher Engineering Mathematics.* Khanna Publisher, 1998.

[74] Cornelius Bode: *Entwicklung und Vergleich unterschiedlicher Rechenmodelle zur Bestimmung von Wirbelstromverlusten in den Permanentmagneten eines Antriebes.* Diplomarbeit, Technical University Braunschweig, 2008.